SUSTAINABLE WATER ENGINEERING

Smart and Emerging Technologies

SUSTAINABLE WATER ENGINEERING

Smart and Emerging Technologies

Edited by

Ali Pourhashemi, PhD

Kian Hariri Asli

Ann Rose Abraham, PhD

A. K. Haghi, PhD

AAP | APPLE ACADEMIC PRESS

First edition published 2025

Apple Academic Press Inc.
1265 Goldenrod Circle, NE,
Palm Bay, FL 32905 USA

760 Laurentian Drive, Unit 19,
Burlington, ON L7N 0A4, CANADA

CRC Press
2385 NW Executive Center Drive,
Suite 320, Boca Raton FL 33431

4 Park Square, Milton Park,
Abingdon, Oxon, OX14 4RN UK

Library and Archives Canada Cataloguing in Publication

CIP data on file with Canada Library and Archives

Library of Congress Cataloging-in-Publication Data

CIP data on file with US Library of Congress

ISBN: 978-1-77491-571-4 (hbk)
ISBN: 978-1-77491-572-1 (pbk)
ISBN: 978-1-00346-859-2 (ebk)

About the Editors

Ali Pourhashemi, PhD
Professor of Chemical and Biochemical Engineering,
Christian Brothers University, Memphis, Tennessee, USA

Ali Pourhashemi, PhD, is a Professor in the Chemical and Biochemical Engineering Department at Christian Brothers University, Memphis, Tennessee, USA. He is involved in teaching, research, and supervising industrial internship programs. Dr. Pourhashemi's primary teaching areas include process design, heat transfer, fluid mechanics, ChE thermodynamics, and unit operations laboratory courses. He has authored several books and research publications in the field of chemical engineering, materials sciences, and related areas, and has been involved in various packaging projects.

Kian Hariri Asli
Research Scholar, Department of Electronic Engineering,
University of Rome "Tor Vergata," Rome, Italy

Kian Hariri Asli is a Research Scholar in the Department of Electronic Engineering, University of Rome "Tor Vergata," Rome, Italy. He has expertise in electronic and control engineering. He works for the development of energy-saving by networked sensors, fuzzy logic, artificial intelligence (AI), and Internet of Things (IoT) based on geospatial information systems (GIS).

Ann Rose Abraham, PhD
Assistant Professor, Sacred Heart College (Autonomous), Thevara, Kochi, Kerala, India

Ann Rose Abraham, PhD, is currently an Assistant Professor at the Department of Physics, Sacred Heart College (Autonomous), Thevara, Kochi, Kerala, India. She has expertise in the field of condensed matter physics, nanomagnetism, multiferroics, and polymeric nanocomposites. She has research experience in various reputed national institutes, including Bose Institute, Kolkata, India; SAHA Institute of Nuclear Physics, Kolkata, India; UGC-DAE CSR Centre, Kolkata, India. She has collaborated with various international laboratories. She is a recipient of Young Researcher award in physics and best paper awards—2020 and 2021. She served as an assistant professor and examiner in the Department of Basic Sciences, Amal Jyothi College of Engineering, under APJ Abdul Kalam Technological University, Kerala, India. Dr. Abraham is a frequent speaker at national and international conferences. She has a good number of publications to her credit in many peer-reviewed high-impact journals of international repute. She has authored many book chapters and edited more than 10 books with Taylor and Francis, Elsevier, etc.

A. K. Haghi, PhD
Research Associate, Department of Chemistry, University of Coimbra, Portugal

A. K. Haghi, PhD, is a retired professor and has written, co-written, edited, or co-edited more than 1000 publications, including books, book chapters, and papers in refereed journals with over 3800 citations and h-index of 32, according to the Google Scholar database. He is currently a research associate at the University of Coimbra, Portugal. Professor Haghi has received several grants, consulted for several major corporations, and is a frequent speaker in national and international conferences. He is a Founder and former Editor-in-Chief of the *International Journal of Chemoinformatics and Chemical Engineering* and *Polymers Research Journal.* Professor Haghi has served as an editorial board member of many international journals. He has served as a member of the Canadian Research and Development Center of Sciences and Cultures. He has supervised several PhD and MSc theses at the University of Guilan (UG) and co-supervised international doctoral projects. Professor Haghi holds a BSc in urban and environmental engineering from the University of North Carolina (USA) and holds two MSc degrees, one in mechanical engineering from North Carolina State University (USA) and another one in applied mechanics, acoustics, and materials from the Université de Technologie de Compiègne (France). He was awarded a PhD in engineering sciences at Université de Franche-Comté (France). He is a persistent reviewer of leading international journals.

Contents

Contributors

Kian Hariri Asli
Department of Electronic Engineering, University of Rome "Tor Vergata," Rome, Italy

P. K. Binsi
ICAR—Central Institute of Fisheries Technology, Matsyapuri, Willington Island, Cochin, India

Mohammad Boroomand
Department of Water and Wastewater Engineering, Shahid Beheshti University, Iran

Bhaskar Gaonkar
Department of Chemistry and Biochemistry, School of Sciences, Jain University (Deemed to be University), Bangalore, Karnataka, India

Gagik Badalians Gholikandi
Department of Water and Wastewater Engineering, Shahid Beheshti University, Iran

S. J. Laly
ICAR—Central Institute of Fisheries Technology, Matsyapuri, Willington Island, Cochin, India

R. Roopashree
Department of Chemistry and Biochemistry, School of Sciences, Jain University (Deemed to be University), Bangalore, Karnataka, India

Negar Saraei
Department of Civil and Environmental Engineering, Lamar University, Beaumont, Texas, USA

V. Swarupa
Department of Chemistry and Biochemistry, School of Sciences, Jain University (Deemed to be University), Bangalore, Karnataka, India

Abbreviations

AMR	antimicrobial resistance
BOD	biological oxygen demand
BPA	bisphenol A
CDs	carbon dots
CFD	computational fluid dynamics
CIP	ciprofloxacin
CR	Congo red
CST	capillary suction time
DNS	direct numerical simulation
DS	dry solids
FDM	finite difference method
FEM	finite element method
FVM	finite volume method
GIS	geographic information system
HA	hydroxyapatite
HB	hydroxybutyrate
HV	hydroxyvalerate
IoT	Internet of Things
LES	large eddy simulation
MW	microwave
MWCNTs	multiwalled carbon nanotubes
NOR	norfloxacin
NRW	non-revenue water
OPP	oxidation reduction potential
OPPs	organophosphorus pesticides
OTC	oxytetracycline
PCF	pressure correction factor
PHA	polyhydroxyalkanoates
PHB	polybetahydroxybutyric
PLA	polylactic acid
PLC	program logic control
PP	polypropylene
RANS	Reynolds-averaged Navier–Stokes
RS	remote sensing

RSM	response surface method
SVI	sludge volume index
SBR	sequencing batch reactors
SCP	single-cell protein
SGS	sub-grid-scale
SRF	specific resistance filtration
SS	solution sample
SVI	sludge volumetric index
TCA	tricarboxylic acid
TCP	tricalcium phosphate
UFW	unaccounted for water
VFA	volatile fatty acid

Preface

Water shortage, municipal population growth, and urban infrastructure degeneration are affecting water security around the world.

This advanced research-oriented volume provides knowledge and guidance for scientists, designers, postgraduate students, researchers, and engineers who are actively involved in the field of water sustainability and are working on water security issues. Additionally, it is also a comprehensive reference book for a broad audience of practitioners and decision makers on new advances in sustainable water management.

The subjects covered in this book include a wide spectrum of water supply and demand, water resources management, and operation and maintenance of water distribution systems. This multidisciplinary reference volume covers sustainability issues from the perspective of integrated water management.

This book also covers interesting topics ranging from water sustainability to water security and safety. It includes novel smart technologies and their industrial applications. To help readers to better understand the subject matter, regional case studies are reported. It is also shown that the application of smart water technologies can help improve services used for water and wastewater services.

Novel smart technologies presented in this valuable reference book provide more efficient methods for the distribution of limited water resources while adding remarkable flexibility to the water system.

CHAPTER 1

Networked Sensors, Remote Sensing (RS), Internet of Things (IoT), and CFD for Water System: Innovations and Challenges in Sustainable Water Management

NEGAR SARAEI

Department of Civil and Environmental Engineering, Lamar University, Beaumont, Texas, USA

ABSTRACT

This work aimed to improve the advanced techniques for the safety and reliability working of storage tanks by improvement of remote sensing (RS) facilities installation equipped with sensors, modems, data loggers, Internet of Things (IoT), and GEO-database intercommunication process. This work, also studied on the one-, two-, and three-baffle cases, effects on eddy currents and stagnation points in reservoir. The results showed they were created in the first region, but no eddy current was formed in the reservoir. The results showed in this case of one-baffle use, eddy current formed in the second region. The flow velocity in the corners has been reduced so much that in the corners of the second zone, stagnation points were formed and the age and time of water retention in these points increased. The option of using two cases of baffle plates was considered. In this case, more eddy currents were generated. As a result, in this case the situation was better than before.

Sustainable Water Engineering: Smart and Emerging Technologies. Ali Pourhashemi, Kian Hariri Asli, Ann Rose Abraham, & A. K. Haghi (Eds.)

1.1 INTRODUCTION

The water supply network of the cities has tasks, the most important of which is the supply of drinking water to the people of the city. These tasks include the supply of water needed by sanitary facilities such as bathrooms, toilets, the supply of water needed by small and large factories and various workshops, water supply necessary to irrigate green spaces and wash streets, and the supply of water needed by firefighting organizations during fires. These tasks also include the components of the water supply network such as tanks, pipes, valves, and pumps.

An urban water supply network must be able to fulfill the mentioned water tasks and requirements in terms of quality (physical and chemical properties of water) and quantity (flow rate and water pressure) according to the existing standards. The supply of required water should be possible in the worst time and place conditions and emergency situations.

Water storage tanks play an important role in the water supply system. Their most important use is to store water in the water supply system for future use or use in times of crisis. They are often installed in high areas, or they may be built on towers that are higher than the highest elevation code of consumption (water head means the equivalent height, a column of water at that height provides the required pressure). For example, water pressure with a head of 30 m is equivalent to a pump that can pump water up to a height of 30 m to meet the needs of the water supply network. They are actually the factor of creating balance between water-producing systems as well as demand. We know that every system has the highest efficiency at its working point, and when it works with a capacity less or more than its capacity, its efficiency is noticeably reduced. The same applies to water purification systems. If the system works with a lower or higher capacity than its design capacity, the efficiency of the system will decrease. On the other hand, demand control is not the hand of the supply side. That is, for some reasons, the demand may suddenly increase during the day and night.

Therefore, it is possible to limit the performance of the treatment plants by analyzing the demand. For example, in the hours when the demand is high, the system works with a higher power and in the hours when the demand is low the system works with a lower power. It was mentioned as the reason and other reasons. Therefore, storing water in low-demand hours and using it in high-demand hours will be a problem. Therefore, the presence of water storage tanks in the water supply system is very vital and guarantees sustainability in the system. Therefore, it is necessary to keep the level of water stored in these tanks at a high level in the water supply network. This is due to the fact that the water level inside the tank must be kept at a certain level and also due to the fact that these tanks have an inlet at the top of the tank and an outlet in the wall facing the location of the inlet and at the bottom of the tank. So the water mixing operation inside the tank may not be done well, and as a result, part of the water will remain in the corner or corner of the tank for a longer period of time than the defined standards, which will lead to water pollution in that area. Also, another factor that can lead to an increase in the retention time of water in the tank or parts of the tank is the geometry of the tank. For example, tanks in which the ratio of the height to the base of the tank is high often increase the retention time.[1] Also, small changes in the water level of the tank lead to an increase in the retention time of the water. The use of weak mixers is also proposed as a solution in order to reduce the water retention time in a corner.[2] But this method can be considered an active method that requires equipment and external energy source to be implemented, so it increases initial investment costs as well as current costs and repair costs. But in the passive method, it does not require an external energy source, and the retention time can be reduced by adjusting the geometric parameters such as the shape of the tank and its dimensions. Since 1960, the method of computational fluid dynamics has been proposed and used to solve and simulate fluid flow.[3]

The main goal of this research is to use computational fluid dynamics to find the effect of different parameters on the retention time of water in the corners of the tank.

Due to the necessity of increasing the quality of water in today's modern society as well as the development of numerical calculation tools, in the last decade much research has been conducted in different parts of the world on how to reduce the retention time or the age of water in the water supply network and especially the existing reservoirs. Increasing the age of water causes the oxidation of disinfectants in water such as chlorine or ozone and causes harmful biological activities in water.[4] These activities severely reduce the quality of water, so it is necessary to take all the necessary measures to reduce the age of water and increase the quality of water.[5] One of the available ways is to improve the hydraulic condition of reservoirs, which is discussed in this research.

1.1.1 WATER STORAGE TANKS

Tanks are generally designed for raw water storage, clean water storage, distribution storage, decontamination, pressure supply, pressure reduction, collection, upstream pumping, or a combination of these.

The horizontal sections of tanks are generally square, rectangular, and in some cases circular and rarely, depending on the location and size of the land, in an unknown shape that is a combination of several rectangular squares. In tanks with a large volume (5000 cubic meters and above), in terms of water movement in the tank, water retention time, etc., the walls of the water flow should be designed to the level of T.W.L. so that the water path from the inlet pipe to the outlet pipe is in a certain direction.

In order to repair, wash the tank, and not cut off the water of the covered distribution network and possible subsequent tanks, the space inside the tank is divided into two or more independent parts, each part can be used as an independent and complete tank.

Reservoirs are classified into two categories, elevated reservoirs and ground reservoirs, according to their location relative to the ground level.

Elevated water tanks are tanks that are built on framed or solid foundations to create proper pressure in the network (due to the absence of natural high ground in the covered area). Under normal conditions, the volume of these tanks is limited and determined based on the need. The examples of elevated water tanks built on framed or integral bases are shown in Figures 1.1–1.4.

FIGURE 1.1 Reinforced concert base for elevated water tank.

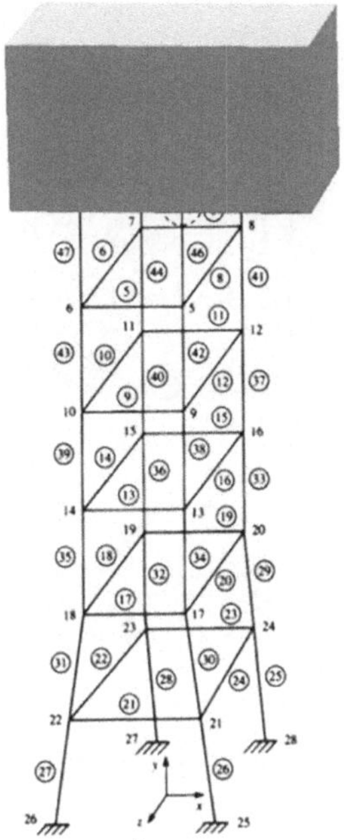

FIGURE 1.2 Elevated water tank: cubic shape.

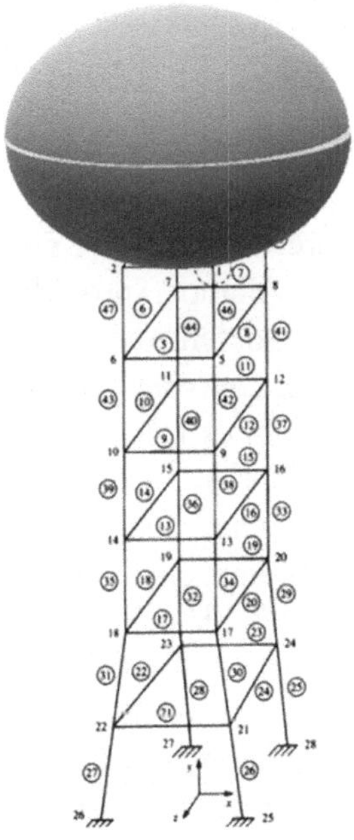

FIGURE 1.3 Elevated water tank: spherical shape.

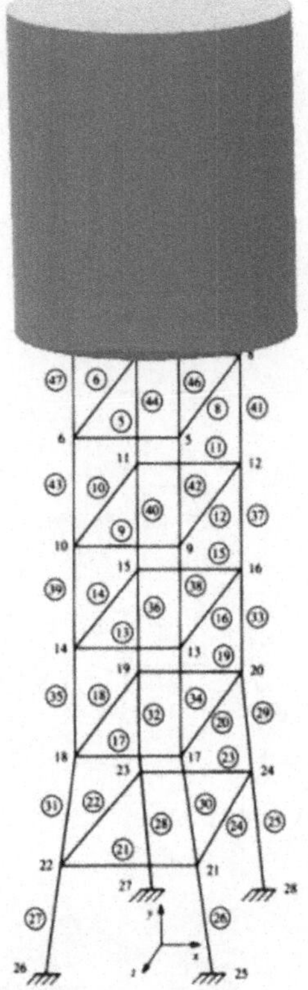

FIGURE 1.4 Elevated water tank: cylindrical shape.

Ground concrete tanks are used for storing purified or sanitary water, balancing pressure, pumping pumps, providing network pressure, etc. According to the rules of passive defense as well as technical and architectural considerations, ground concrete tanks may be built as buried, semiburied, or exposed which are shown in Figures 1.5–1.9.

Ground concrete tanks are buried tanks that are established at a suitable depth and then suitable soil is poured behind the walls and also on their roof and the soil behind the walls is compacted. In addition to the merits of camouflage, such tanks also provide suitable conditions in terms of heat exchange.

In semiburied ground concrete tanks, the excavation is done only up to a part of the height of the wall. In terms of camouflage, temperature changes, as well as the expansion and contraction of the roof slab, such tanks have inappropriate conditions and are not recommended for use in the urban drinking water network due to the nonactive defense criteria.

FIGURE 1.5 Ground water tank: cylindrical shape.

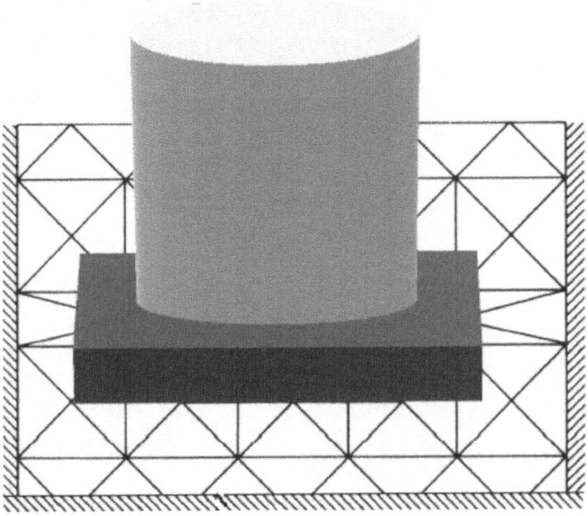

FIGURE 1.6 Reinforced concert base of water tank: cylindrical shape.

In visible concrete tanks, they are usually made visible in terms of landscape architecture or symbolism, and in accordance with the surrounding environment, to organize the image and urban, historical, and touristic landscapes. Figures 1.7–1.9 show the examples of buried, semiburied, and exposed ground concrete tanks. The factors that affect the choice of the type of underground reservoirs in terms of buried, semiburied, or visible include passive defense regulations, underground water level, climatic conditions, topography, geo-technics, land limitations, the desired amount, the slope of the land, or the need to organize views and landscapes. In order to comply with the rules of passive defense and provide the necessary measures to deal with frostbite, buried ground tanks should be used for drinking water. If it is not possible to bury the tanks for such centers, the tank should be studied according to the inactive defense committee's opinion and according to the relevant regulations. In such conditions, due to the rules of passive defense, it is recommended not to use tanks whose geometric shape has sharp corners.

FIGURE 1.7 Underground water tank: cylindrical shape.

FIGURE 1.8 Semi-underground water tank: cylindrical shape.

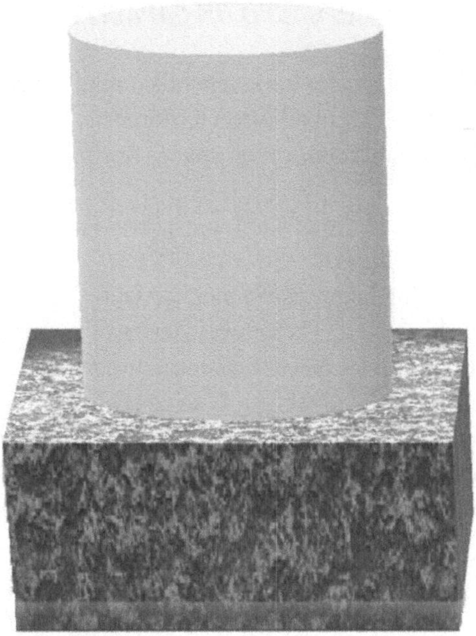

FIGURE 1.9 Ground water tank: cylindrical shape.

Reservoirs are used in water supply systems such as catchments, transmission lines, treatment plants, and distribution networks for storage, disinfection, pressure relief, pressure supply, well water collection, and a combination of them, in cities and villages.

Ground concrete water tanks are usually built in two shapes: cylinders and rectangular cubes, although they can be built in any beautiful and appropriate geometric shape. In general, cylindrical tanks are superior to rectangular cubic tanks from the technical point of view and passive defense considerations. In areas where the soil load factor is suitable, bowl tanks are also a suitable option for large volumes. From the point of view of exploitation and passive defense considerations, tanks are usually considered twins. For example, instead of building a 5000 cubic meter tank, two 2500 cubic meter tanks are built next to each other or attached to each other. The dimensions of the tanks depend on the characteristics of the land and environmental conditions, but if there are no special limitations, the most suitable geometric dimensions for the rectangular cubic tanks are obtained in terms of economy in the conditions where the ratio of the length to the width of the tank is 3:2.[6–10] In twin cylinder tanks, the most suitable case is to use two independent tanks with equal volume.

1.1.2 DIVISION OF TANKS BASED ON CAPACITY

The volume of the potable water tank depends on the consumption of the flat area in the target year, which in addition to domestic consumption, administrative and general, commercial, green space, firefighting, industrial, etc. is meant in its estimation.

It is possible that there are ground and elevated reservoirs with smaller capacity in cities and villages.

The tank capacity for large distribution networks may be more than one hundred thousand cubic meters. Therefore, in terms of using and maintaining the tank space, in most of the tanks, they are designed in several parts that are adjacent and connected to each other, and each part can be used alone or out of the system.

Therefore, according to the mentioned cases, the division of the tanks according to the capacity of each part of the tank in terms of the possibility of proper exploitation and maintenance of them according to the capacity was described.

1.1.3 TYPES OF TANKS IN TERMS OF CONSTRUCTION MATERIALS

In general, the construction materials for the foundation, wall, and roof of the tank according to the cost, construction time, useful life, capacity, and strength against the internal and external forces of water pressure such as the reaction of the ground and surrounding walls, overheads such as soil cover on the tank and around it, the underground water around the tank, etc. are different from each other and each of them may be superior to the other in one or more mentioned factors.

Currently, the construction materials in the existing tanks are as follows:

- Reinforced concrete
- Steel
- Building materials with beams or block beams
- Fiberglass
- Iron and glass

1.1.4 TYPES OF TANKS ACCORDING TO THE TYPE OF INTERNAL SURFACES

The internal surfaces of the tanks are lined with building materials, sand, and cement, and in some cases, they are tiled. In many tanks, the internal surfaces are covered with resin or chemicals, or they are uncoated depending on the material. In some cases, standard polymer coating is used for drinking water tanks.

1.1.5 TYPES OF TANKS IN TERMS OF APPLICATION

1.1.5.1 RAW WATER STORAGE TANK

This tank is designed to store and balance the raw water of treatment plants, which is often open and is not the subject of this research.

1.1.5.2 STORAGE TANK FOR DISINFECTION

The tank that is used to create water contact time with chlorine or other disinfectants is called the proximity tank. Its capacity is based on the required retention time, the capacity of the treatment plant, the capacity of the pumps, the total capacity of the produced water, and the needs determined. These types of tanks in treatment plants are known as clean water tanks.

1.1.5.3 STORAGE TANK

This type of tank is for storing water. This reservoir may be multi-purpose; in other words, it may include one or more other users.

1.1.5.4 PRESSURE TANK

This type of tank is for providing proper pressure in the distribution network to which water is transferred with or without a pump.

1.1.5.5 PRESSURE RELIEF TANK

This tank is designed to reduce the pressure to zero in the path of the transmission lines where the elevation difference is high and at certain points.

1.1.5.6 COLLECTION TANK

This tank is usually designed to collect the produced water of the well complex in one area and transfer their water to the point of consumption, with or without a pump house as needed. In fact, in terms of chlorination, these reservoirs are water extracted from wells or nearby reservoirs.

1.1.6 SOME DESIGN REQUIREMENTS

Basically, the design and construction of reservoirs is based on detailed studies related to supply, demand, type of supply, consumption, water quality, climate, economic and income conditions, welfare level, culture, development, and many other effective factors for the project period. But the followings are the most important things that must be considered in the design.

1.1.7 ESTIMATION OF WATER CONSUMPTION

The amount of water for domestic, public, commercial and industrial use, green space, firefighting, and livestock farming in villages, etc. is estimated for the target year, which is usually equivalent to a useful life period of the structure.

1.1.8 STABLE WATER DISTRIBUTION

To ensure that the water is not cut off, the volume of the tank should be determined.

The amount of consumption in the downstream distribution network, the possible transfer of water to other reservoirs, compliance with the excess volume for the possible cut-off time of the inlet pipes, and no overflow of water taking into account the optimized diagram of inputs and outputs based on the economic justification of construction and operation provided.

1.1.9 STREAM-LINE FLOW

Inside the ground tank parts of groups two and three create a stream-line flow of water and use all the useful volume of the tank and target dead spots from the mouth of the inlet pipe to the mouth of the outlet pipe in order to prevent stagnation. Part of the water inside the tank, circulation walls are expected.

1.1.10 ROUND CORNER

The exposed corners of the tank walls are designed to be rounded to prevent water from stagnating in that area.

1.1.11 INSPECTION WINDOWS

Often, in the designs, access and entry hatches to the tank with dimensions of at least 100 × 80 cm and at a height of 80 cm above the top of the tank roof with a lock for complete protection of the tank are provided, and in the two walls of the vent valve with the stairs and lighting of the water level of the tank are designed.

1.1.12 TANK COVER

Due to the fact that water has destructive effects on steel and concrete over time, the inside and outside of metal and concrete tanks should be coated with resin or chemicals that do not adversely affect the quality of water. Most of the coatings that are used are[11–17]:

- Epoxy resin
- Vinyl
- Glazing
- Coal tar

1.1.13 AGE OF WATER FLOWING FROM THE TANK

The design of drinking water reservoirs should be in such a way that the age of the water is reduced to the minimum acceptable from the point of view of health, economy, and engineering. The circulation of water in the hydraulic system of the tank prevents the creation of stagnant areas. If the inlet of the tank is designed correctly, the inlet water should provide the necessary mixing in the tank, provided that this process travels a long way in the tank and the incoming water can provide a turbulent flow (based on the conducted research and previous studies,[18–22] the ratio of delivered flow rate (Q) to the diameter (d) of the inlet pipe).

It should also be noted that the difference in the temperature of the inlet water and the aqueous space of the tank may cause thermal layering, which does not result in effective mixing. There is a possibility of stratification of water in a tank with a large inlet. To prevent it, it is necessary to increase the amount of incoming water so that the increase in the volume of incoming water leads to more turbulent flow and as a result mixing of the flow and loss of layers. The critical difference between the temperature of the incoming water and the water in the tank (T in degrees Celsius) that leads to thermal stratification in the tank can be calculated from eq (1.1).

$$T = \frac{CQ^2}{D^3 H^2} \tag{1.1}$$

In the above relationship, *C* is a dimensionless coefficient that depends on the shape of the water inlet, the buoyancy model, and the diameter of the tank and is often calculated experimentally. *Q* is the inlet water flow rate in liters per minute. Also, *H* and *D* express the height of the tank in meters and the diameter of the tank in millimeters.[23]

1.1.14 GROUND TANK SYSTEMS

1.1.14.1 ROOF STRUCTURAL SYSTEM

Tanks can be built with or without a roof. Potable water tanks must be covered and impenetrable. Roofless tanks are usually used in cases such as sedimentation ponds, swimming pools, and agricultural water storage. One of the following systems can be chosen for the roof of the tanks:

- Mushroom flat slab
- Tir-Dal

According to the characteristics of the structures and also the required reliability conditions for the roof of the tanks, the mushroom flat slab type system is the most suitable and the most common system. Although, in some cases, a double-sided slab system with environmental support beams is also used for urban drinking water tanks. In case of using a double-sided slab system with peripheral bearing beams, the technical specifications necessary for reliability and also the performance of the slab as a rigid diaphragm should be studied.

1.1.15 CONNECTING THE ROOF TO THE WALL IN GROUND TANKS

Usually, the connection of the roof to the wall in semiburied tanks is considered free in order to reduce the effect of thermal length changes, and in buried tanks, to prevent the penetration of underground water, it is considered integral. The connection between the external environment and the internal environment of the tanks should be completely cut off by using suitable seals at the place of connection between the roof and the wall. The examples of the connection systems between the roof and the wall of ground tanks are shown in Figure 1.10.

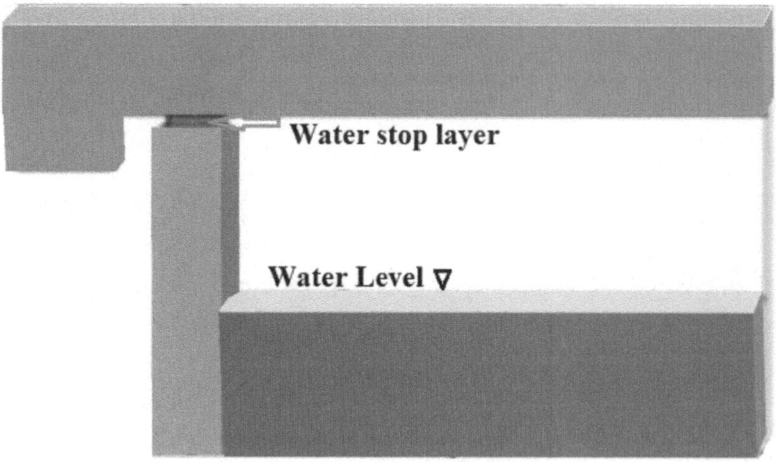

FIGURE 1.10 Connecting the roof to the wall of ground tanks.

It should be noted that in Figure 1.10 the longitudinal and transverse changes of the roof of concrete tanks due to temperature changes are shown in an enlarged form. These changes in buried tanks are much smaller than unburied tanks.

1.1.16 WALL STRUCTURAL SYSTEM IN GROUND RECTANGULAR CUBE TANKS

The wall systems of ground concrete tanks are considered integral or nonintegral according to the presence of vertical expansion joints in the wall, according to one of the following situations.

a. *One-cloth system*

- Two-sided slab, supported on three edges and free on the upper edge (the supporting condition may be hinged or clamped), in this case, the walls are subjected to longitudinal tension in addition to the bending anchor.
- Two-sided slab, resting on four edges (the supporting position may be hinged or clamped), in this case, the walls are subjected to longitudinal tension and vertical compression in addition to the bending anchor.

b. *Nonintegrated system*

- One-sided vertical slab, clamped at the bottom and free at the top
- One-sided vertical slab, clamped at the bottom and connected to the ceiling at the top, which is usually assumed to be a hinged wall and ceiling connection.

In the nonintegrated system, the walls and floor have expansion joints (Figure 1.11).

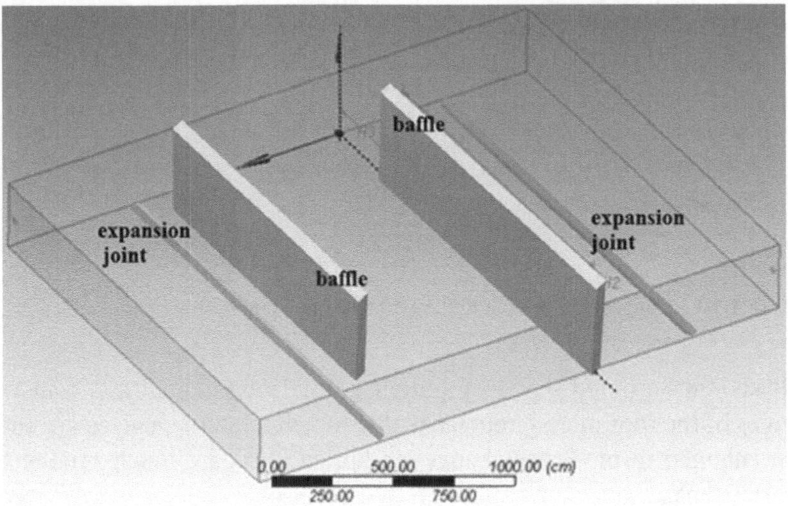

FIGURE 1.11 Baffles and expansion joints of ground tanks.

1.1.17 WALL STRUCTURAL SYSTEM IN COLUMN TANKS

Wall systems of cylindrical concrete tanks are considered integral or nonintegral according to the presence of vertical expansion joints in the wall, according to one of the following situations.

a. Integrated system

In this system, there is no vertical expansion joint in the wall, and in the cylindrical shell, in addition to the annular tension, there is a bending moment with different clamping conditions at the bottom and top of the shell.

b. Nonintegrated system

In this system, there is a vertical expansion joint in the wall and the following two situations occur:

- One-sided vertical slab, clamped at the bottom and free at the top.
- One-sided vertical slab, clamped at the bottom and connected to the ceiling at the top (which is usually assumed to connect the wall to the hinged ceiling).

Since the most important advantage of cylindrical tanks is the creation of annular tension, and in the case of expansion joints, annular tension is not created, usually a nonintegral system is not used for cylindrical concrete tanks.

1.1.18 METHODOLOGY AND RESEARCH OBJECTIVES

In this research, the main goal is to find the number and locations of the inlet and outlet of the water storage tank in such a way that the retention time of water in all parts of the tank is as uniform and minimal as possible, and as a result, the quality of the outlet water reaches the desired standards. Also, another goal is to place the separators (baffles) in the tank in terms of arrangement and number, which is done with the aim of uniformity and minimizing the water retention time in the tank.

To achieve these goals, the dynamic simulation of water flow inside the tank will be done using computational fluid dynamics methods, and the water flow pattern and, as a result, the retention time will be calculated for different input and output states and separators.

Tanks in the water distribution network are responsible for storing water for use in emergency situations, firefighting and controlling fluctuations in consumption. Several decisions such as the location, size, and type of tank operation in the design of the tank in water distribution networks must be made.[24] Three types of storage in the distribution system include fire storage, volume balancing, and emergency storage for the required tank volume. All tanks must be filled and emptied within their operating range without using the emergency volume.[25] The main advantage of choosing the location for new tanks in the first place is to provide the reliability of the network at the time of pipe failure and to create the appropriate pressure in the network nodes. In the second step, feeding the network from different directions helps to minimize the diameter of the pipes and as a result to reduce the age of water in the network and reduce the retention time of water in the tanks.[26] The size of the tanks is determined based on certain parameters such as demand balance, emergency storage, and firefighting needs.[27] The higher capacity of the tank causes more demand of the network and thus increases its reliability. On the other hand, increasing the capacity of tanks increases the costs and water retention time, and then reduces the quality of water in the tank, and designers must maintain a balance between these two opposing goals.[28] Unlike the reliability of water distribution systems, which is the universal definition, there is no acceptance for it.[29] Reliability of storage tanks can be defined based on its ability to store water for consumers. If the tank is completely empty, it will fail. Its reliability can be defined based on the probability of tank failure. Various failure parameters, including the number of failures, length of failure time, and tank failure time, can be used.[30]

In general, very few modeling studies have been conducted to investigate the flow inside the water storage tanks as well as the quality of the stored water.[31] In fact, in these modeling, the tank should be considered a nonuniform storage tank because in some corners the fluid may have a high retention time and in some places not.

Mau and his colleagues[32] in 1995 presented an explicit model for simulating water flow and water quality inside a storage tank. In their study, they defined three different behaviors (different scenarios) for the storage tank and examined the water quality in each case. They wrote mass transfer, mixing, and kinetic reaction relations for reservoirs. Then the obtained equations were solved numerically. They used two different numerical methods to solve the equations. Then, the amount of chloride and fluoride injected into the tank was calculated in certain points which were called sensitive points by using numerical solution and measured experimentally. Then, using experimental data, they investigated the validity of the presented numerical method. The numerical data were in good agreement with the experimental data. Also, they calculated the concentration of fluoride and chloride for different tank volumes at a given point and presented them under graphs. In fact, their article can be considered a primer on mathematical modeling of water storage tanks with an appropriate accuracy. Therefore, their article is important from this point of view.

In 2000, Rossman[33–34] considered four different models for the storage tank and investigated the water flow and water quality in each case. He stated that the water quality inside the storage tank is highly dependent on the geometry of the flow and consequently on the geometry of the tank. They show that the flow entering into the tank leads to complete mixing of the flow inside the tank or not depends on the ratio of the momentum of the incoming flow to the buoyancy of the incoming flow. A similar situation exists for output currents. They showed that there are only limited geometries in which complete mixing can take place, which, of course, is related to the volume of the tank and the ratio of its height to its base.

He investigated the behavior of a number of rectangular storage tanks. He examined the pattern of the flow field and the age of water and showed that in tanks with horizontal inlets, better mixing occurs than vertical inlets at a high level. As the ratio of length to width increases in tanks with a vertical inlet, the flow characteristics move toward creeping flows at a faster rate.

Mahmood and his colleagues[35] in 2005, using the computational fluid dynamics model, investigated different samples of existing storage tanks in the drinking water supply system of Virginia. They compared the obtained numerical results with experimental data obtained from measurements at different points of the storage tank. Their results showed that the computational fluid dynamics model has a good accuracy in simulating the fluid flow inside the tank. And it can be used to determine the proper location of tank inlets and outlets. First, they examined the speed of the flow entering the tank and the position of the pipe entering the tank on the mixing rate of the tank. For this purpose, simulation was carried out in a tank with a storage volume of 3.8 mega-liters. They assumed that a flow with a volumetric flow rate of 63 L per second enters the tank at a speed of 29 m per minute through a vertical inlet with a diameter of 400 mm. They considered the fluid temperature inside the tank to be 20°C and the inlet fluid temperature to be 18°C. Then the concentration distribution inside the tank was obtained after 60 min using computational fluid dynamics.

Then, the diameter of the Verdi tube was considered to be equal to 1200 mm, and as a result, the speed of the inlet jet was equal to 3.7 m/min. In this case, other boundary conditions were considered similar to the previous case. Then the concentration distribution inside the tank was obtained after 60 min.

They concluded that the presence of flow with a higher inlet velocity will lead to better mixing in the tank.

Another factor that could affect the amount of mixing is the type of input orientation. For this purpose, they investigated two types of vertical and horizontal input orientation. For this purpose, they considered a storage tank with a volume of 15 mega-liters with an input volumetric flow rate of 126 L per second. The diameter of the vertical tube was 900 mm and the jet inlet speed was 11 m/min. In this case, the conditions were also assumed to be isothermal. That is, the temperature of the tank was assumed to be equal to the temperature of the inlet fluid and equal to 20°C. Then, the concentration inside the tank was calculated for different orientations of the Verdi tube.

They concluded that the vertical inlet direction will lead to better mixing in the tank.

They stated that the temperature difference between the flow of fluid entering the tank and the fluid stored in the tank can be an effective factor on mixing. For this purpose, the effect of this parameter was investigated numerically using computational fluid dynamics. For this purpose, they considered a water storage tank with a volume of 15 mega-liters with a vertical inlet. They considered the mass flow rate of the inlet fluid to be equal to 126 L per second and the diameter of the Verdi tube to be equal to 900 mm. As a result, the inlet jet speed was equal to 11 m/min. Then they once assumed that the temperature of the tank water is equal to 25°C and the temperature of the inlet water is equal to 24°C and once again the temperatures of the inlet water and the stored water are the same.

Finally, they validated the obtained numerical results with laboratory data measured by themselves.

Also, in 2006, Matinez[69] developed a method based on computational fluid dynamics to simulate two-dimensional flow inside a rectangular tank. They considered a rectangular tank with dimensions of 20 × 25 m, height of 3 m, and volume of 5000 cubic meters, which has an inlet, an outlet, and a place for injecting injectable. The flow rate at the entrance is considered to be equal to 50 L per second, and for the system to work in stable mode; accordingly, the flow rate at the outlet must be equal to 50 L per second. Also, they considered the diameter of the inlet and outlet pipes to be equal to 250 mm, assuming that the injected materials are added by a third pipe with a diameter of 40 mm. They also assumed that the inlet velocity of the inlet pipe and the injection pipe is equal to 1 m per second. You can see a view of the geometry studied by them in Figures 1.2–1.5.

In 2006, Farmani et al.[36] investigated the application of multi-objective evolutionary algorithm in identifying the interaction between total cost, network resilience (hydraulic reliability), and water age (water quality) in water distribution systems. They tested this on a real network like Anytown. They considered cost minimization, including the investment costs of pipes, tanks, and the current cost of consumed energy, as one of the optimization goals. So far, their goal is to optimize the overall cost, water quality, and reliability of the network, and the variables of the problem design include adding sight pipelines, replacing the previous pipelines or making changes in their structure, building new storage tanks or increasing the storage capacity of the previous tanks has also changed the working hours of the pump. Also, from their point of view, the total cost includes the operating cost of the pump, the cost of installing a new storage tank, and increasing the capacity of the existing tanks, as well as the cost of piping. Finally, they presented a water distribution network as an optimal state.

They also considered the resilience presented by Todini[37] in 2000 as their goal to increase the availability of water when pipes break. Their goal is to design a water distribution system with high reliability, and overall cost has been low. Therefore, they used the multi-objective optimization algorithm and defined total cost, reliability, and water quality as objective functions. They defined three types of distribution systems and calculated the total cost for each and compared them.

In 2006, Prasad[38] designed a water distribution system with storage tanks. He explained a method for designing the water distribution network along with the correct selection of various elements such as pipes, pumps, and storage tanks. He used the genetic algorithm in steady state to optimize the water distribution system. His objective function was to minimize the total cost. In his research, total cost is defined as the sum of initial investment costs and energy costs. His goal was to determine the size of the pipes, the required pump, and the pump's operating time, and to choose the number of tanks and the capacity of the tanks so that the objective function is optimized. Finally, they used to minimize the water retention time in the network as a quality parameter for the third objective function. Their results showed that the design of the network in this way leads to a network with high capability, so that the network obtained has optimal pipes and the best arrangement for tanks in high consumption points.

In 2007, Marek et al[39] investigated the jet flow inside the tank by solving the averaged Navier–Stokes unsteady state equations. They stated that the mixing process inside the water storage tanks is an essential parameter in the quality of the stored water. They stated that various factors are effective on the mixing inside the tank, including the power of the inlet jet speed, nozzle diameter, and nozzle inlet angle. Their goal was to determine the optimal value for the nozzle angle and the diameter of the nozzle entering the tank. They compared the results of their simulations with the available experimental data and showed that the presented results were in good agreement with each other. They considered a cylindrical tank with one inlet and one outlet in three dimensions and investigated it numerically and experimentally.

First, they calculated the flow lines inside the tank for the inlet nozzle angle of 45° and the inlet jet velocity of 4.4 m/s.

Then they investigated the influence of the angle and diameter of the inlet nozzle on the flow field. For this purpose, they obtained the contour of the velocity field in three cases. In the first case, the angle of the inlet nozzle was considered to be equal to 45° and the diameter of the inlet nozzle was equal to 21 mm, and as a result, the speed of the inlet jet was equal to 2.1 m/s. In the second case, the angle of the nozzle was considered to be equal to 45° according to the previous case, but in this case, the diameter of the inlet nozzle was considered to be equal to 8 mm, and as a result, the speed of the inlet jet was equal to 4.4 m/s. And in the third case, the angle of the inlet nozzle was considered to be equal to 90°, and the diameter of the inlet nozzle was assumed to be equal to 8 mm. The velocity contour along the middle plane y = 0.

They found that the most optimal mode is obtained when the nozzle angle is equal to 45° and the inlet nozzle diameter is equal to 8 mm. As a result, they investigated the concentration of materials injected into the tank under these conditions to see how the distribution of the injected materials would be at different times. The concentration of the substance injected inside the tank was obtained after the lapse of 10 s, 20 s, and 30 s. In fact, the concentration of the substance injected into the tank can be considered a measure of the retention time of the water inside the tank. That is, the areas that do not change after the injection of hand sanitizer are the same areas where the water stays and the quality of the water decreases.

1.2 METHODS

In the new global view, water is considered a basic human need. Although water is one of the renewable resources, its amount is limited. In the present study, the behavior of hydraulic parameters including velocity, water retention time, and also leakage from the reservoir in relation to simulation of the position of the partition walls (baffles) in the concrete ground tank have been investigated. The main purpose of this study is to determine the impact of reservoir design specifications on hydraulic parameters and the possibility of water leakage and economic losses. Therefore, determining the dimensions, number, and optimal location for the location of the partition walls was also considered in this study and as a research hypothesis, options one and two of the partition wall and their effect on the field and flow lines were investigated.

The main question of the present work is how many baffle plates by creating eddy currents in different areas can reduce the stagnation points and thus reduce the water retention time. So using it compared to the previous cases can have a greater comparative advantage.

The volume of the reservoir was equal to 5000 cubic meters. The diameter of the inlet and outlet pipes was 25 cm and the diameter of the injection pipe of the plug was 4 cm and the inlet and outlet water flow rate was measured at about 50 L per second. These data were detected by the remote sensing flowmeters. Therefore, the volume of the reservoir was determined to be 5000 cubic meters. Therefore, if the tank is full, about 8 h will be enough to supply water, which is almost in accordance with the standard. The average water retention time in the reservoir was 8 h. The water inlet velocity as well as the injection rate of the plug was achieved to reach a flow rate of 50 L per second at about 1 m/s. The length of the pipes is equal to 4 m and its length was 3 m outside the reservoir and 1 m inside the reservoir based on the criteria and criteria for design and calculation of groundwater reservoirs. Also, inlet, outlet, and injection pipes were assumed to be in the middle height of the tank and the distance of the inlet pipe from the side wall was equal to 0.5 m. The distance of the outlet pipe from its side wall was assumed to be equal to 0.5 m.

For the inlet pipe, the inlet velocity condition was used. For the walls of reservoir, the non-slip condition was used. The value of the wall roughness was assumed to be 0.1 mm. For the free surface, the symmetry boundary condition was used. This condition increased the number of required iterations and decreased the degree of convergence in the problem. Then the zero shear stress boundary condition was used. This boundary condition somewhat improved the solution criteria. Therefore, it can be concluded that this boundary condition is more appropriate. The output pressure limit condition was used for the reservoir output. The species transport model was also used to investigate the concentration of injected particles.

In this case, no partition walls (baffles) were intended for the reservoir, so this case was considered the base. As mentioned earlier, the computational model is based on solving the mass conservation of mass equation and the Navier–Stokes equations for fluid motion. Therefore, if the flow is turbulent, the turbulent flow equations are added to the basic equations. Numerical methods used in the computational model are: finite element method (FEM), finite volume method (FVM), finite difference method (FDM). Therefore, finite volume method (FVM) is used more in modeling incompressible flows. The CFD software output based on finite volume method (FVM) was used for modeling.

In the present study, the problem was meshed by using the CFD software and the meshing of the element. For roughness, 0.1 mm was considered for the inner wall of the tank and it was assumed that the water level inside the tank was constant. Unstructured mesh type was used to mesh the volume inside the tank and for this purpose, mesh with hexagonal (octagonal) cells was used. Mesh with different dimensions and types were examined and the duration and number of repetitions required for the convergence of the problem in each case were compared with each other to use the best type of mesh. Finally, the view of meshing criteria embedded in CFD software was used (Figure 1.12).

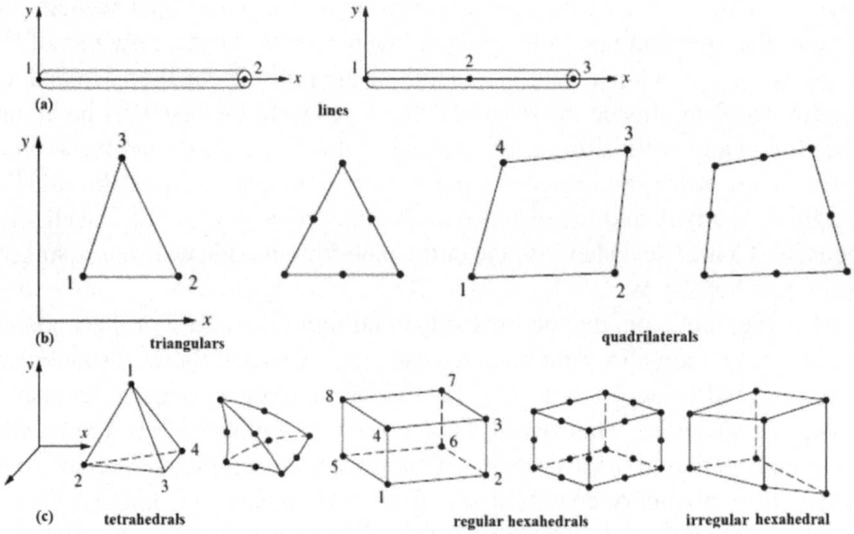

FIGURE 1.12 Meshing criteria embedded in CFD software: (a) one-dimentional element, (b) two-dimentional element, and (c) three-dimentional element.

1.2.1 *FORMULATION*

The computational model is based on solving the conversion of mass equation and the Navier–Stokes equation for fluid motion. Therefore, if the flow is turbulent, the turbulent flow equations are added to the basic eqs 1.2–1.24:

Conservation of mass:

$$\frac{\partial \bar{\rho}}{\partial t}+\frac{\partial}{\partial x_i}\left(\overline{\rho u_i}\right)+\frac{\partial}{\partial x_i}\left(\overline{\rho' u_i'}\right)=0 \tag{1.2}$$

Conservation of momentum:

$$\rho\frac{D\bar{V}}{Dt}+\rho\frac{\partial}{\partial x_i}\left(\rho\overline{u_i' u_j'}\right)=\rho g-\nabla\bar{P}+\mu\nabla^2\bar{V} \tag{1.3}$$

In the above relation, μ kinematic viscosity, $\bar{P}$ pressure vector, $\bar{V}$ velocity vector, g gravity acceleration, $\frac{D\bar{V}}{Dt}$ express the ordinary derivative of velocity vector with respect to time.

$$\rho\frac{D\bar{V}}{Dt}=\rho g-\nabla\bar{P}+\nabla\tau_{ij} \tag{1.4}$$

In the above relation, τ_{ij} expresses the tension tensor, which is defined as follows:

$$\tau_{ij}=\tau_{ij,Laminar}+\tau_{ij,Turbulent}=\left[\mu\left(\frac{\partial u_i}{\partial x_j}+\frac{\partial u_j}{\partial x_i}\right)\right]-\rho\overline{u_i' u_j'} \tag{1.5}$$

In the above relation, $\tau_{ij,Laminar}$ and $\tau_{ij,Turbulent}$ express the laminar current stress tensor and the turbulent current stress tensor, respectively.

Conservation of energy:

$$\rho C_P\frac{D\bar{T}}{Dt}=-\frac{\partial}{\partial x_i}\left[-k\frac{\partial\bar{T}}{\partial x_i}+\rho C_P\overline{u_i' T'}\right]+\frac{\mu}{2}\overline{\left[\frac{\partial\bar{u}_i}{\partial x_i}+\frac{\partial u_i'}{\partial x_j}+\frac{\partial\bar{u}_j}{\partial x_i}+\frac{\partial u_j'}{\partial x_i}\right]^2} \tag{1.6}$$

In the above relation, C_P expresses the specific heat capacity at constant pressure, $\bar{T}$ temperature, and k the thermal conductivity.

The kinetic energy equation of perturbation

$$K=\frac{1}{2}\overline{u_i' u_i'}=\frac{1}{2}(\overline{u'u'}+\overline{v'v'}+\overline{w'w'}) \tag{1.7}$$

$$\frac{DK}{Dt}=\frac{\partial}{\partial x_i}\overline{[u_i'(\frac{1}{2}u_j' u_j'+\frac{P'}{\rho})]}-\overline{u_i' u_i'}\frac{\partial\bar{u}_j}{\partial x_i}+\frac{\partial}{\partial x_i}\overline{[vu_j'(\frac{\partial u_i'}{\partial x_j}+\frac{\partial u_j'}{\partial x_i})]}$$
$$\overline{v\frac{\partial u_j'}{\partial x_i}\left(\frac{\partial u_i'}{\partial x_j}+\frac{\partial u_j'}{\partial x_i}\right)} \tag{1.8}$$

In the above relation, K expresses the kinetic energy of turbulence and v expresses dynamic viscosity.

The intensity TI perturbation is related to the kinetic energy and the mean Uref as reference current velocity as eqs 1.3–1.8.

$$\mathrm{TI} = \frac{[\frac{1}{3}\left(\overline{u^2} + \overline{v^2} + \overline{w^2}\right)]^{0.5}}{U_{ref}} = \frac{[\frac{2}{3}K]^{0.5}}{U_{ref}} \tag{1.9}$$

Conservation of mass:

$$\frac{\partial \rho}{\partial t} + \frac{\partial}{\partial x_i}(\rho u_i) = 0 \tag{1.10}$$

$$\frac{\partial \rho u_i}{\partial t} + \frac{\partial}{\partial x_j}\left(\rho u_i u_j\right) = \frac{\partial P'}{\partial x_j} + \frac{\partial}{\partial x_i}\left[\mu_{ref}\left(\frac{\partial u_i}{\partial x_j} + \frac{\partial u_j}{\partial x_i}\right)\right] + S_M \tag{1.11}$$

SM is sum of physical forces, μ_{eff} effective viscosity and P′. The pressure is defined by the following relation:

$$P' = P + \frac{2}{3}\rho k + \frac{2}{3}\mu_{eff}\frac{\partial u_k}{\partial x_k} \tag{1.12}$$

According to the following relation, kinetic energy depends on the flow of energy.

$$\mu_t = C_\mu \rho \frac{k^2}{\varepsilon} \tag{1.13}$$

The value of *k* and ε C_μ is a fixed number.

The ordinary differential equations of kinetic energy transfer and perturbation energy drop rates are obtained as follows:

$$\frac{\partial \rho k}{\partial t} + \frac{\partial}{\partial x_j}\left(\rho u_j k\right) = \frac{\partial}{\partial x_j}\left[\left(\mu + \frac{\mu_t}{\sigma_k}\right)\frac{\partial k}{\partial x_j}\right] + P_k - \rho\varepsilon + P_{kb} \tag{1.14}$$

$$\frac{\partial \rho \varepsilon}{\partial t} + \frac{\partial}{\partial x_j}\left(\rho u_j \varepsilon\right) = \frac{\partial}{\partial x_j}\left[\left(\mu + \frac{\mu_t}{\sigma_\varepsilon}\right)\frac{\partial \varepsilon}{\partial x_j}\right] + \frac{\varepsilon}{k}(C_{\varepsilon 1}P_k - C_{\varepsilon 2}\rho\varepsilon + C_{\varepsilon 1}P_{\varepsilon b} \tag{1.15}$$

$C_{\varepsilon 1}$, $C_{\varepsilon 2}$, σ_k σ_ε, and σ_ε are constant. P_{kb} and $P_{\varepsilon b}$ exert the influence of the Buoyancy forces. Pk produces turbulence due to viscous forces.

$$Pk = \mu_t\left(\frac{\partial u_j}{\partial x_j} + \frac{\partial u_i}{\partial x_i}\right)\frac{\partial u_i}{\partial x_j} - \frac{2}{3}\frac{\partial u_k}{\partial x_k}\left(3\mu_t\frac{\partial u_k}{\partial x_k} + \rho k\right) \tag{1.16}$$

It is assumed that viscosity or perturbation $k-\omega$ is related by the following kinetic energy relationship of perturbation and perturbation frequency:

$$\mu_t = \rho \frac{k}{\omega} \tag{1.17}$$

The values of k and ω are obtained from the differential equations of kinetic energy transfer and perturbation frequency:

$$\frac{\partial \rho k}{\partial t} + \frac{\partial}{\partial x_j}\left(\rho u_j k\right) = \frac{\partial}{\partial x_j}\left[\left(\mu + \frac{\mu_t}{\sigma_k}\right)\frac{\partial k}{\partial x_j}\right] + P_k - \beta' \rho k\omega + P_{kb} \text{ (1.18)} \tag{1.18}$$

$$\frac{\partial \rho \omega}{\partial t} + \frac{\partial}{\partial x_j}\left(\rho u_j \omega\right) = \frac{\partial}{\partial x_j}\left[\left(\mu + \frac{\mu_t}{\sigma_\omega}\right)\frac{\partial \omega}{\partial x_j}\right] + \alpha \frac{\omega}{k} P_k - \beta \rho \omega^2 + P_{\omega b} \tag{1.19}$$

$$\beta' = 0.09, \alpha = \frac{5}{9}, \beta = 0.075, \sigma_k = 2, \sigma_\omega = 2 \tag{1.20}$$

$$\mathrm{F2} = \tanh(arg_2^4) \tag{1.21}$$

$$\mathrm{arg1} = \min[\max(\frac{\sqrt{k}}{0.09\omega y}, \frac{500\mu}{\rho\omega y^2}), \frac{4\rho k}{CD_{k\omega}\sigma_{\omega,2} y^2}] \tag{1.22}$$

$$\mathrm{arg2} = \min[\max(\frac{\sqrt{k}}{0.09\omega y}, \frac{500\mu}{\rho\omega y^2})] \tag{1.23}$$

$$CD_{k\omega} = \max[2\rho \frac{1}{\sigma_{\omega,2}} \frac{1}{\omega} \frac{\partial k}{\partial x_j} \frac{\partial \omega}{\partial x_j}, 10^{-10}] \tag{1.24}$$

1.3 RESULTS AND DISCUSSION

This work investigated the maximum water retention time due to the effect of dimension, numbers, and optimal location of baffles as efective factors for water leakage. As a result, the effect of angle and baffle walls on the flow field in reservoirs was calculated based on the Geospatial Information System (GIS). According to the GIS, the relationship classes between the spatial data and nonspatial data were defined for the scaled model. In the field test process, the spatial data and nonspatial data were collected and transmitted to Program Logic Control (PLC). The PLC was used as a tool for spatial data intercommunication of water system elements for water leakage investigation. To reduce the water leakage, the hydraulic model for the analysis of water loss was defined in compliance with GIS based on the ArcGIS-ArcMap, Fluent, and Gambit software.

The purpose of this work is to simulate the mixing processes and the performance of the reservoir. At the first, mixing parameters will be checked.

Many parameters have been introduced in the maximum research for mixing processes. These parameters can be generally divided into two categories: Eullerian parameters and Lagrangian parameters. The parameters of the first group (Eulerian parameters) are related to the cases that are used for mixing and are mainly based on hydrodynamic parameters. These parameters indirectly affect the mixing process and fluid flow inside the reservoir. The reservoir input is the main variable in this area. The other parameters in the first group can be the intensity of turbulence, retention time, and velocity gradient. In the second group, items such as concentration histogram were included.

The velocity contour on the middle plate of the tank is shown in Figures 1.13–1.16. The volume fraction contour of the injected material was obtained.

Due to the fact that the main purpose of this study is to investigate the retention time and velocity profiles in the tank to determine the stagnation points and optimize the geometry, in order to avoid the complexity of the problem in the next models, the injection of additives is omitted. Flow lines are checked to determine stationary points and points with high retention times. The numbers of baffles are considered in Figures 1.13–1.16.

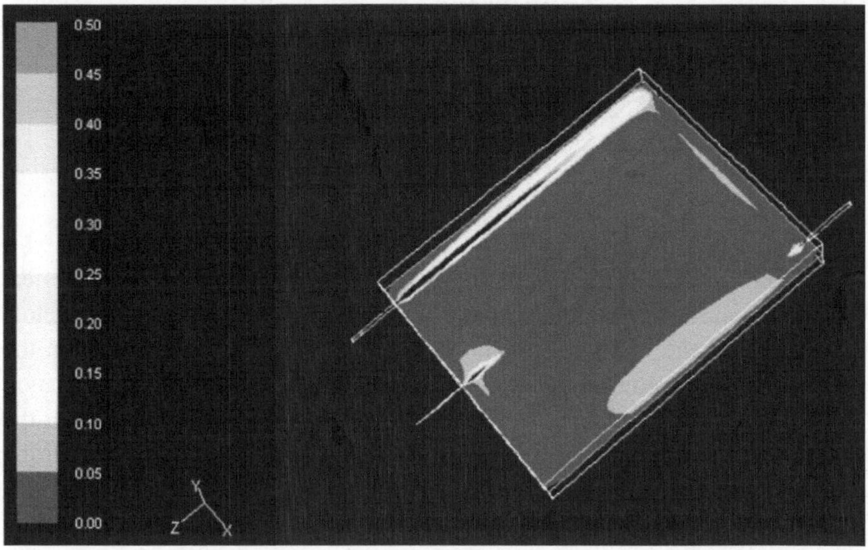

FIGURE 1.13 Speed meter at the middle height of the water storage tank.

In the present study, it was shown that if additives are injected into the tank, the concentration of these materials in the corners will be greatly reduced, and as a result, this type of tank will have many disadvantages.

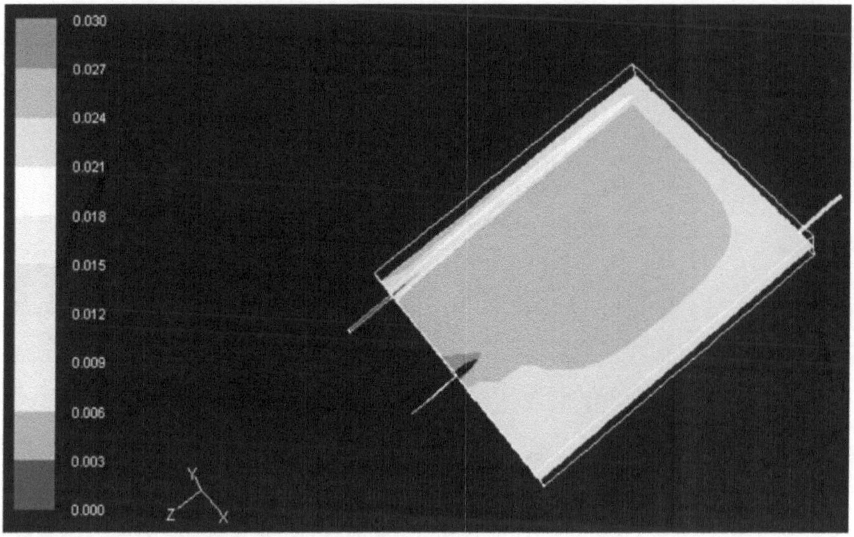

FIGURE 1.14 Volumetric fraction of the injected material.

To solve this problem, baffle should be used. Therefore, the effect of using baffles on the flow field was studied.

Using the baffle leads to the formation of eddy currents and as a result a point of stillness is formed in the corners of the tank. Therefore, water age and retention time increase in these areas. By comparing the flow lines and the different modes under consideration, it can be concluded that the tank-related option with three baffles cases mounted completely vertically in front of the inlet current is the best expected mode.

To avoid the complexity of the problem, in the next models, the injection of additives was omitted and the velocity vector and flow lines were considered to determine the stationary points and points with long retention time. In this case, two baffles (Figures 1.15–1.16) were considered.

In this regard, the velocity contour, flow lines, and velocity vector are shown in Figures 1.16.

As can be seen in Figures 1.15 and 1.16, the addition of two baffle plates leads to the emergence of three eddy currents in the three areas created by the two baffle plates. This leads to better flow mixing and reduced retention time in the areas where the vortex is formed. If the areas formed by the baffle plates from the left are called one, two, and three, respectively, it can be seen in Figures 1.15 and 1.16. In the lower corner of the second area and especially in the upper corner of the third area, stagnation points are formed at which the fluid retention time increases slightly.

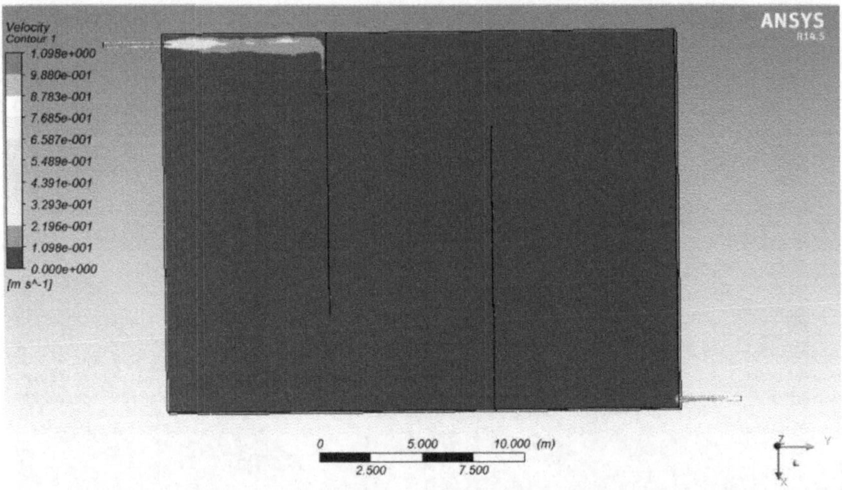

FIGURE 1.15 Speed meter line for three-dimensional water storage tank consisting of two baffles at a distance of 8 m from each other.

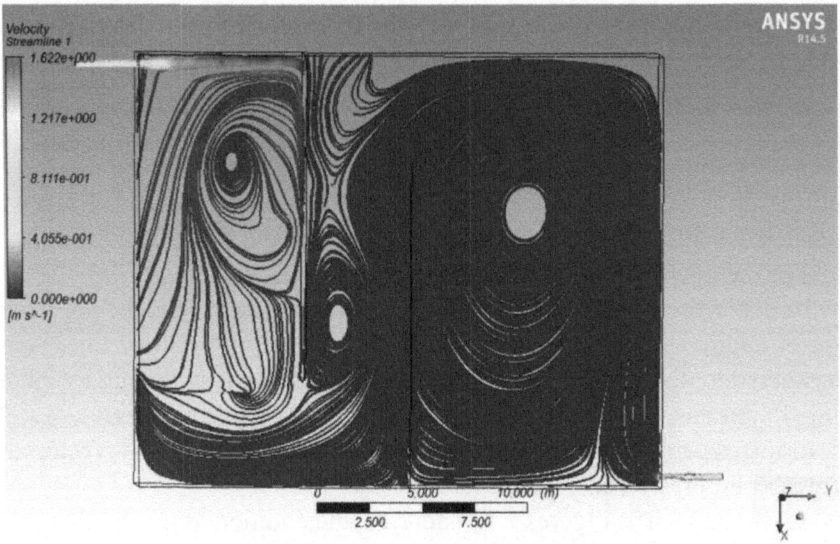

FIGURE 1.16 Flow lines for a three-dimensional water storage tank consisting of two baffles at a distance of 5 m from each other.

In the following, the effect of the distance between the baffle plates on the flow field and flow lines is investigated. For this purpose, once the distance between the plates is 10 m and again 5 m.[19–30]

1.3.1 RESEARCH RESULTS OF ANOTHER GROUP OF SCIENTISTS[69]

In this work, the effect of the distance of the baffles from each other on the flow field was investigated. It was shown that the reduction of baffles leads to a weakening of the vortex strength, especially in the second and third regions. Increasing the number of baffle plates is strongly effective on increasing the eddy currents formed.

Matinez[69] also developed a method based on the computational fluid dynamics (CFD) in 2006 to simulate two-dimensional flow inside a rectangular tank. He considered a rectangular tank with dimensions of 20 × 25 m, height of 3 m, and volume of 1500 cubic meters, which had an inlet, an outlet, and a place for injecting injection materials. The flow rate at the input is set at 50 L per second, and for the system to work in a steady state, the flow rate at the output must be equal to 50 L/s. They also considered the diameter of the inlet and outlet pipes to be 250 mm and assumed that the injectable material was added by a third pipe with a diameter of 40 mm. They also assumed that the inlet velocities of the inlet pipe and the injection pipe were equal to 1 m/s. A view of the handhelds studied by them can be seen in Figure 1.17.

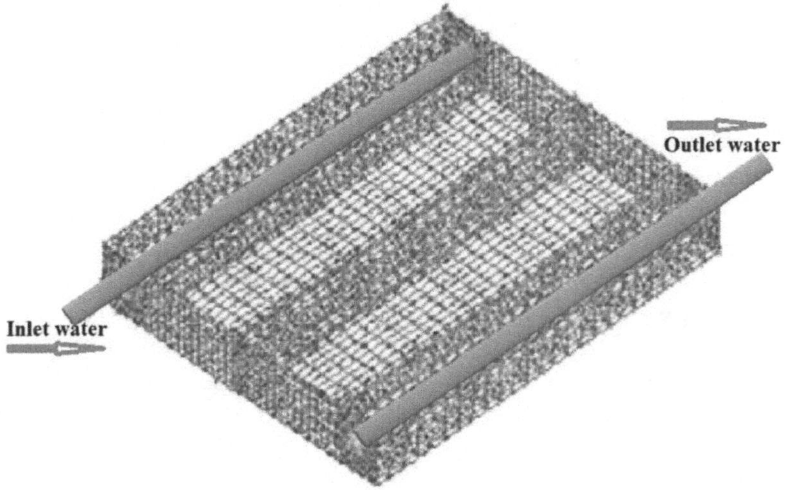

FIGURE 1.17 Flow lines for a three-dimensional water storage tank.

1.4 CONCLUSION

This work showed the addition of two baffle plates led to the emergence of three eddy currents in the three areas created by the two baffle plates. This referred to better flow mixing and reduced retention time in the areas where the vortex was formed. If the areas formed by the baffle plates from the left were called one, two, and three, respectively, in the lower corner of the second area and especially in the upper corner of the third area, stagnation points were formed at which the fluid retention time increases slightly.

This can lead to water corruption and increases the risk of water leakage in the reservoirs. Another factor that can increase the water retention time in the reservoirs is the shape and geometry of the reservoirs. The main purpose of this work was to determine the impact of reservoirs design specifications on the probability of water leakage and economic losses resulting from it. In this regard, this work focused on the effects of dimension, numbers, and optimal location of baffles as an efective factor on water leakage.

KEYWORDS

- **Internet of Things**
- **remote sensing**
- **water retention time**
- **baffle**
- **reservoir**
- **water leakage**
- **networked sensors**

REFERENCES

1. Asli, K. H.; Aliyev, S. A. O.; Thomas, S.; Deepu, A. *Handbook of Research for Fluid and Solid Mechanics: Theory, Simulation, and Experiment*; Apple Academic Press; Taylor Francis Group: New York, 2017. https://doi.org/10.1201/9781315365701.
2. Hariri Asli, H.; Arabani, M.; Golpour, Y. Reclaimed Asphalt Pavement (RAP) Based on a Geospatial Information System (GIS). *Slovak J. Civil Eng.* **2020,** *28* (2), 36–42. https://doi.org/10.2478/sjce-2020-0013.
3. NZS 3106. *Design of Concrete Structures for the Storage of Liquids, Cement and Concrete Association of New Zealand*; Standards Association of New Zealand, 2009.
4. *Recommended Standards for Water Works, Great Lakes-Upper Mississippi River Board of State and Provincial Public Health and Environmental Managers*; Health Research Inc., Health Education Services Division: P.O. Box 7126, Albany, NY 12224, 2012 102- SA Water TG.

5. Planning and Infrastructure, Above Ground Circular Concrete Tank, Manager Engineering, 2008. 103-Selected Earthquake Engineering Papers of George W. Housner, American Society of Civil Engineers, 1990.
6. *Specification for Construction of Concrete Reservoirs, Standard Specifications and Drawings Specification for Construction of Concrete Reservoirs*; Gold Coast, Australia, SS 11, 2004.
7. TM 5-809-10, *Seismic Design Guidelines for Essential Buildings*; US Department of the ARMY, Washington, D.C., 1986.
8. USEPA Finished Water Storage Facilities, U.S. Environmental Protection Agency, 2002.
9. Zecchin, A. C.; Maier, H. R.; Simpson, A. R.; Leonard, M.; Nixon, J. B. Parametric Study for an Ant Algorithm Applied to Water Distribution System Optimization. *IEEE Trans. Evol. Comput.* **2005,** *9* (2), 175–191.
10. Walski, T. M. Hydraulic Design of Water Distribution Storage Tanks. In *Water Distribution Systems Handbook*; Mays, L. W., eds.; McGraw-Hill: New York, 2000; Chapter 10.
11. Farmani, R.; Walters, G.; Savic, D. Evolutionary Multi-objective Optimization of the Design and Operation of Water Distribution Network: Total Cost vs. Reliability vs. Water quality. *J. Hydroinform.* **2006,** *8* (3), 165–179.
12. Haarhoff, J.; Van Zyl, J. E.; Nel, D.; Van der Walt, J. J. Water Supply and Water Quality in Rural Areas: Stochastic Analysis of Rural Water Supply Systems. *Water Supply* **2000,** *18* (1), 467–470.
13. Hariri Asli, H. Investigation of the Factors Affecting Pedestrian Accidents in Urban Roundabouts. *Trans. Civil Environ. Eng.* **2022,** *2226,* 1–4. https://doi.org/10.52547/crpase.8.1.2255.
14. Hariri Asli, H.; Arabani, M. Analysis of Strain and Failure of Asphalt Pavement, Computational Research Progress in Applied Science & Engineering. *Trans. Civil Environ. Eng.* **2022,** *8,* 1–11, Article ID: 2250. https://doi.org/10.52547/crpase.8.1.2250.
15. https://en.wikipedia.org/wiki/Computational_physics
16. Maier, H. R.; Simpson, A. R.; Zecchin, A. C.; Foong, W. K.; Phang, K. Y.; Seah, H. Y.; Tan, C. L. In *Ant Colony Optimization for the Design of Water Distribution Systems, Proceedings of World Water and Environmental Resources Congress 2001,* ASCE, The Rosen Plaza Hotel: Orlando, Florida, United States, 2001.
17. Maier, H. R.; Simpson, A. R.; Zecchin, A. C.; Foong, W. K.; Phang, K. Y.; Seah, H. Y.; Tan, C. L. Ant Colony Optimization for Design of Water Distribution Systems. *J. Water Resour. Plan. Manag.* **2003,** *129* (3), 200–209.
18. Walski, T. M.; Chase, D. V.; Savic, D. A.; Grayman, W.; Beckwith, S.; Koelle, E.; *Advanced Water Distribution Modeling and Management*; Haestad Press; Watertown, Conn, 2003.
19. Hashemi, S. S. Optimization of Water Networks With Minimization of Pumping Energy. Master Thesis, University of Tehran; Iran, 2010.
20. Mau, R. E.; Boulos, P. F.; Clark, R. M.; Grayman, W. M.; Tekippe, R. J.; Trusell, R. R. Explicit Mathematical Models of Distribution Storage Water Quality. *ASCE J. Hydraul. Eng.* **1995,** *121* (10), 669–709.
21. Rossmann, L. A. *EPANET 2 User's Manual*; Water Supply and Water Resources Divison. U.S. [Online] 2000. Environmental Protection Agency. Cincinnati, EE.UU. http://www.epa.gov/ORD/NRMRL/wswrd/epanet.html
22. Hariri Asli, H.; Nazari, S. Water Age and Leakage in Reservoirs: Some Computational Aspects and Practical Hints. *Larhyss J.* **2021,** *48,* 151–167. http://larhyss.net/ojs/index.php/larhyss/article/view/808/807.

23. Yeung, H. Modeling of Service Reservoirs. *J. Hydroinform.* **2001,** *3* (3), 165–172.
24. Mahmood, F.; Pimblett, J. G.; Grace, N. O.; Grayman, W. M. Evaluation of Water Mixing Characteristics in Distribution System Storage Tanks. *J. AWWA,* **2005,** *97* (3), 74–88.
25. Hariri Asli, H.; Hozori, A. Non-Revenue Water (NRW) and 3D Hierarchical Model for Landslide. *Larhyss J.* **2021,** *48,* 189–210. http://larhyss.net/ojs/index.php/larhyss/article/view/810/810.
26. Marek, M.; Stoesser, T.; Roberts, P. J. W.; Weitbrecht, V.; Jirka, G. H. *CFD Modeling of Turbulent Jet Mixing in a Water Storage Tank.* XXXII IAHR Congress; Venice (Italy), July 1/6, 2007.
27. Van Zyl, J. E.; Haarhoff, J.; Reliability Analysis of Municipal Storage Reservoirs Using Stochastic Analysis. *J. S. Afr. Inst. Civ. Eng.* **2007,** *49* (3), 27–32.
28. Hariri Asli, K.; Khodaparast Haghi, A.; Hariri Asli, H.; Sabermaash Eshghi, E. Water Hammer Modelling and Simulation by GIS. *Model. Simul. Eng.* **2012,** *704163,* 4. doi:10.1155/2012/704163. http://www.hindawi.com/journals/mse/aip/704163/.
29. Manning, G. P. *Concrete Reservoirs and Tanks*; Concrete Publication, 1972.
30. *Minimum Design Loads for Buildings and Other Structures*; American Society of Civil Engineers, 2010.
31. Nourbakhsh, S. A.; Jaumotte, B. A.; Hirsch, C.; Parizi, H. B. *Turbopumps and Pumping Systems*; Springer, 2008.
32. Mau, R. E.; Boulos, P. F.; Clark, R. M.; Grayman, W. M.; Tekippe, R. J.; Trusell, R. R. Explicit Mathematical Models of Distribution Storage Water Quality. *ASCE J. Hydraul. Eng.* **1995,** *121* (10), 669–709.
33. Rossmann, L. A. *EPANET 2 User's Manual.* Water Supply and Water Resources Divison. U.S. [Online] 2000. Environmental Protection Agency. Cincinnati, EE. UU. http://www.epa.gov/ORD/NRMRL/wswrd/epanet.html
34. Rossmann, L. A.; Grayman, W. M. Scale-model Studies of Mixing in Drinking Water Storage Tanks. *J. Environ. Eng.* **2000,** *125* (8), 755–761.
35. Mahmood, F.; Pimblett, J. G.; Grace, N. O.; Grayman, W. M. Evaluation of Water Mixing Characteristics in Distribution System Storage Tanks. *J. AWWA* **2005,** *97* (3), 74–88.
36. Farmani, R.; Savic, D. A.; Walters, G. A. *The Simultaneous Multi-Objective Optimization of Anytown Pipe Rehabilitation, Tank Sizing, Tank Siting and Pump Operation Schedules*; Critical Transitions in Water and Environmental Resources Management, ASCE, American Society of Civil Engineers: Salt Lake City, 2004; pp 4663–4672.
37. Todini, E. Looped Water Distribution Networks Design Using a Resilience Index Based Heuristic Approach. *Urban Water* **2000,** *2* (3), 115–122.
38. Prasad, T. D. Design of Pumped Water Distribution Networks with Storage. *J. Water Resour. Plan. Manag.* **2010,** *136* (1), 129–132.
39. Marek, M.; Stoesser, T.; Roberts, P. J. W.; Weitbrecht, V.; Jirka, G. H. *CFD Modeling of Turbulent Jet Mixing in a Water Storage Tank*; XXXII IAHR Congress: Venice (Italy), July 1/6, 2007.
40. Munshi, J. A.; *Rectangular Concrete Tanks*; Portland Cement Association, 1998.
41. Munshi, J. A. *Circular Concrete Tanks Without Prestressing*; Portland Cement Association, 1993.
42. Munshi, J. A. *Design of Liquid Containing Concrete Structures for Earthquake Forces*; Portland Cement Association, 2002.

43. NCHRP Report 611. *Seismic Analysis and Design of Retaining Walls, Buried Structures, Slopes, and Embankments*; American Association of State Highway, National Cooperative Highway Research Program, 2008.
44. Nourbakhsh, S. A.; Jaumotte, B. A.; Hirsch, C.; Parizi, H. B. *Turbopumps and Pumping Systems*; Springer, 2008.
45. NSF/ANSI 61. *Drinking Water System Components—Health Effects*; NSF National Standard/ANSI National Standard, 2010.
46. NZS 3106. *Design of Concrete Structures for the Storage of Liquids*; Cement and Concrete Association of New Zealand, Standards Association of New Zealand, 2009.
47. *Recommended Standards for Water Works, Great Lakes-Upper Mississippi River Board of State and Provincial Public Health and Environmental Managers*; Health Research Inc.; Health Education Services Division, P.O.Box 7126, Albany, NY 12224, 2012 102-SA Water TG.
48. Planning and Infrastructure, Above Ground Circular Concrete Tank, Manager Engineering, 2008. 103- Selected Earthquake Engineering Papers of George W. Housner; American Society of Civil Engineers, 1990.
49. *Specification for Construction of Concrete Reservoirs, Standard Specifications and Drawings Specification for Construction of Concrete Reservoirs*, Gold Coast, Australia, SS 11, 2004.
50. TM 5-809-10. *Seismic Design Guidelines for Essential Buildings*; US Department of the ARMY; Washington, D.C.; 1986.
51. USEPA. *Finished Water Storage Facilities*; U.S. Environmental Protection Agency, 2002.
52. Zecchin, A. C.; Maier, H. R.; Simpson, A. R.; Leonard, M.; Nixon, J. B.; Parametric Study for an Ant Algorithm Applied to Water Distribution System Optimization. *IEEE Transc. Evol. Comput.* **2005,** *9* (2), 175–191.
53. Walski, T. M.; Hydraulic Design of Water Distribution Storage Tanks. In *Water Distribution Systems Handbook*; Mays, L. W., Ed.; McGraw-Hill: New York, 2000; Chapter 10.
54. Farmani, R.; Walters, G.; Savic, D. Evolutionary Multi-objective Optimization of the Design and Operation of Water Distribution Network: total cost vs. Reliability vs. Water Quality. *J. Hydroinform.* **2006,** *8* (3), 165–179.
55. Farmani, R.; Savic, D. A.; Walters, G. A.; *The Simultaneous Multi-Objective Optimization of Anytown Pipe Rehabilitation, Tank Sizing, Tank Siting and Pump Operation Schedules*; Critical Transitions in Water and Environmental Resources Management, ASCE, American Society of Civil Engineers, Salt Lake City, 2004; pp 4663–4672.
56. Haarhoff, J.; Van Zyl, J. E.; Nel, D.; Van der Walt, J. J. Water Supply and Water Quality in Rural Areas: Stochastic Analysis of Rural Water Supply Systems. *Water Supply* **2000,** *18* (1), 467–470.
57. Maier, H. R.; Simpson, A. R.; Zecchin, A. C.; Foong, W. K.; Phang, K. Y.; Seah, H. Y.; Tan, C. L. In *Ant Colony Optimization for the Design of Water Distribution Systems*, Proceedings of World Water and Environmental Resources Congress 2001, ASCE, The Rosen Plaza Hotel, Orlando, Florida, United States, 2001.
58. Maier, H. R.; Simpson, A. R.; Zecchin, A. C.; Foong, W. K.; Phang, K. Y.; Seah, H. Y.; Tan, C. L. Ant Colony Optimization for Design of Water Distribution Systems. *J. Water Res. Plan. Manag.* **2003,** *129* (3), 200–209.
59. Walski, T. M.; Chase, D. V.; Savic, D. A.; Grayman, W.; Beckwith, S.; Koelle, E.; *Advanced Water Distribution Modeling and Management*; Haestad Press; Watertown, Conn, 2003.

60. Rossmann, L. A. *EPANET 2 User's Manual*; Water Supply and Water Resources Divison. U.S. 2000. Environmental Protection Agency. Cincinnati, EE.UU. http://www.epa.gov/ORD/NRMRL/wswrd/epanet.html
61. Rossmann, L. A.; Grayman, W. M. Scale-Model Studies of Mixing in Drinking Water Storage Tanks. *J. Environ. Eng.* **2000,** *125* (8), 755–761.
62. Yeung, H. Modelling of Service Reservoirs. *J. Hydroinform.* **2001,** *3* (3), 165–172.
63. Van Zyl, J. E.; Haarhoff, J.; Reliability Analysis of Municipal Storage Reservoirs Using Stochastic Analysis. *J. S. Afr. Inst. Civil Eng.* **2007,** *49* (3), 27–32.
64. Gualtieri, C. *Analysis of Flow and Concentration Fields in a Baffled Circular Storage Tank*; XXXIII IAHR Congress: Vancouver, Canada, 2009; pp 4384–4392.
65. MacPherson, G.; Cripps, S.; de Schaetzen, W. *Storage and Pumping Capacity Analysis Using Hydraulic Modeling*, Conference Proceeding Paper, A. A. Balkema Publishers, Water Management Challenges in Global Changes, CCWI: London, England, 2007, pp 439–446.
66. Teixeira, E. C.; Siqueira, R. N. Performance Assessment of Hydraulic Efficiency Indexes. *J. Environ. Eng.* **2008,** *134* (10), 851–859.
67. Sturman, R.; Wiggins, S. Eulerian Indicators for Predicting and Optimizing Mixing Quality. *N. J. Phys. 11* (6), 75031.
68. Metcalf & Eddy, Inc. *Wastewater Engineering Treatment and Reuse*. McGraw-Hill, 1995. Patwardhan, A. W. CFD Modeling of Jet Mixed Tanks. *Chem. Eng. Sci.* **2002,** *57,* 1307–1318.
69. Martínez-Solano, F. J.; López-Jiménez, P. A.; Iglesias-Rey, P.; Pérez, R. "Utilización de la fluidodinámica computacional (CFD) como herramienta para la determinación de parámetros de los depósitos en modelos de calidad del agua" (in spanish). V Congreso Ibérico sobre gestión y planificación del agua. Universidad de Algarbe: Faro, Portugal, 2006.
70. Hariri Asli, K.; Hariri Asli, K. Isolated Pressure Zones Based on GIS as a Solution for Water Network Problems. *J. Water Pract. Technol.* **2023,** *17* (9), in press.
71. Asli, H. H. Reclaimed Asphalt Pavement (RAP) Based on Geospatial Information System (GIS) and Networked Sensors Modelling, New Advances in Material Technologies Experimental Characterizations, Theoretical Modeling, and Field Practices; in Press by the Apple Academic Press, Inc., Hard [Online] (2023). ISBN: 9781774914847, USA, Canada, https://www.appleacademicpress.com/new-advances-in-material-technologies-experimental-characterizations-theoretical-modeling-and-field-practices/9781774914847
72. Asli, H. H. Applications of Networked Sensors and the Internet of Things (IoT) for Wastewater Treatment, Sustainable Water Treatment and Ecosystem Protection Strategies; in Press by the Apple Academic Press, Inc., Hard [Online] 2023. ISBN: 9781774915189, USA, Canada, https://www.appleacademicpress.com/sustainable-water-treatment-and-ecosystem-protection-strategies-/9781774915189
73. Asli, H. H. Applications of Networked Sensors and the Internet of Things (IoT) for Water Treatment, Sustainable Water Treatment and Ecosystem Protection Strategies; in Press by the Apple Academic Press, Inc., Hard [Online] 2023. ISBN: 9781774915189, USA, Canada, https://www.appleacademicpress.com/sustainable-water-treatment-and-ecosystem-protection-strategies-/9781774915189
74. Asli, H. H. Modeling of Corrosion for Water System by Networked Sensors and the Internet of Things (IoT) in Compliance with Geography Information System (GIS), Sustainable Water Treatment and Ecosystem Protection Strategies, in Press by the Apple

Academic Press, Inc.: Hard [Online] 2023. ISBN: 9781774915189, USA, Canada, https://www.appleacademicpress.com/sustainable-water-treatment-and-ecosystem-protection-strategies-/9781774915189

75. Asli, H. H.; Pourhashemi, A.; Abraham, A. R.; Haghi, A. K. New Advances in Material Technologies Experimental Characterizations, Theoretical Modeling, and Field Practices, in Press by the Apple Academic Press, Inc.: Hard [Online] 2023. ISBN: 9781774914847, USA, Canada, https://www.appleacademicpress.com/new-advances-in-material-technologies-experimental-characterizations-theoretical-modeling-and-field-practices/9781774914847
76. Asli, H. H.; Tatrishvili, T.; Abraham, A. R.; Haghi, A. K. *Sustainable Water Treatment and Ecosystem Protection Strategies*, in Press by the Apple Academic Press, Inc., Hard [Online] 2023. ISBN: 9781774915189, USA, Canada, https://www.appleacademicpress.com/sustainable-water-treatment-and-ecosystem-protection-strategies-/9781774915189

CHAPTER 2

Using Rice Husk and Sludge-Rice Husk Mixture Ash for Application of Waste-Activated Sludge Thickening and Sustainable Dewatering Process

MOHAMMAD BOROOMAND[1], NEGAR SARAEI[2], and GAGIK BADALIANS GHOLIKANDI[1]

[1]*Department of Water and Wastewater Engineering, Shahid Beheshti University, Iran*

[2]*Department of Civil and Environmental Engineering, Lamar University, Beaumont, Texas, USA*

ABSTRACT

Finding the best sludge treatment process necessitates conducting a case study of the sludge treatment experiment because each treatment plant's sludge properties are distinct. The current study's objective was to identify the most practical and cost-effective method for treating biological sludge. Two varieties of rice husk, one with a poor yield and the other with a high yield, were researched. For the conditioning procedure, rice husk powder and sludge ash were used to improve the thickening, dehydration, and digestibility of the sludge. Using a RSM test design, SRF, and CST test, the effectiveness of conditioners on the dewatering process was assessed. Calculations were made to determine the quantity of integral elements, such as sludge pH and conditioner injection dose. Low (L) yield rice husk powder, high (H) yield rice husk powder, and mixing of sludge ash with rice husk have optimal pH of 7, 6.3, and 7, respectively, in the process of biological

Sustainable Water Engineering: Smart and Emerging Technologies. Ali Pourhashemi, Kian Hariri Asli, Ann Rose Abraham, & A. K. Haghi (Eds.)

biodegradation sludge. The optimal injection dose of L, H rice husk powder and mixture was equivalent to 0.3, 0.25, and 1 g/gDS, respectively. The blend of sludge ash and rice husk powder appears to perform better than the two rice husk samples, based on the SVI index data. Generally, the usage of a mixture of rice husk powder and sludge ash has a better performance in the digestion process.

2.1 INTRODUCTION

Nowadays, sludge management is a significant part of solid and liquid waste management in most countries as a fundamental activity. US Environmental Protection Agency has reported that more than one million tons of sludge per year are generated in wastewater treatment plants of the United States. The development of conventional biological systems for wastewater treatment leads to the growth of global sludge generation. As a result, sludge management has been one of the most critical environmental issues in the past and future decades.[1] The increase in public environmental awareness and the pressures of environmental organizations highlight the need to find an economical and environmentally friendly solution for sludge management and wastewater treatment problems.[2] To maintain a sustainable environment, hazardous materials must be removed from urban areas.[3,4] Sustainable practices promote the health and vitality of the environment, people, and the economy. According to sustainability, resources are limited and should be used prudently with an eye toward long-term goals and the effects of how they are used.[5,6]

The main purpose of sludge treatment is to achieve quantitative and qualitative specifications, in accordance with the existing criteria for its final disposal or reuse. Sludge treatment and disposal of urban wastewater treatment plants are primarily done to stabilize the sludge and oxidize the organic materials. This issue causes the sludge to be less harmful from a bacteriological point of view, and its purification and reuse can be done more easily due to the reduction in volume and percentage of moisture.[7]

Three main mechanisms of solid particles settling, filtration, compaction, and compression have been used in the usual facilities for sludge thickening and dewatering. The physical and chemical characteristics of sludge, the type of stabilization process, the type of conditioning process, and the type of sludge dewatering process are the most vital factors that affect the mechanical separation of water from sludge.[8] Thickening is an operation with the aim of increasing volume of solids in the sludge by separating a part of the effective water in it. This process helps reduce the volume of the sludge and the capacity of the reactors related to the subsequent purification processes. The amount of chemicals is reduced in the sludge conditioning process and the amount of heat energy in the digester. Another advantage of sludge thickening is the reduction of sludge pumping and transportation costs.[2] It is often difficult to dewater sludge and biological solids containing colloidal particles without conditioning. In addition to the above, some of the sludge conditioning methods for its disinfection, the sludge conditioning starts with processes such as coagulation and flocculation to accelerate the sedimentation process, and optimally increases the weight percentage of sludge solids. The thickening process continues until the dehydration because the coagulation process reduces the electrostatic repulsion forces between sludge solid particles. In the continuation of the flocculation process, it causes the compaction and accumulation of fine colloidal particles and the trapping of solids by increasing the amount of sludge collision. In this way, the specific resistance of sludge filtration is reduced.[9] The appropriate dewatering process among the mentioned methods is selected according to the characteristics of the sludge, the features of the dewatered by-products, and the required space. Among the most common methods of dewatering sludge, the use of belt press filters, pressure press filters, vacuum filters, sludge drying beds, and lagoons is common.[10] Usually, the process of burning sludge is done to reduce the volume and weight, and as a result, to reduce the requirements of sludge disposal, disinfection of sludge and production of sterile ash, destruction of toxic compounds and energy recovery. However, this brings disadvantages such as demanding high investment and operation costs, the need for experienced skilled operators, and air pollution.[11,12]

After wheat, rice occupies the largest area of agricultural land in the world. Disposal of burnt rice husk ash may also be associated with environmental and health problems. According to the results of the investigations, most of the rice husk is made up of silica (20–40%).[13,14] By adding two agricultural wastes, bamboo powder and rice husk powder, Wang et al.[15] demonstrated that sludge dewatering is also facilitated and reduces chemical oxygen demand. Rice husk powder reduces the density of sludge from 1.08 to 0.78. Simultaneous conditioning with bamboo powder and rice husk powder reduces capillary suction time and filtration resistance. Wu et al.[16] investigated the possibility of conditioning and dewatering sludge with rice husk modified by cationic ferric chloride. The results showed that rice husk modified by ferric chloride with a dose of 3 mol/L for 1 h can increase the sludge dewatering ability to the title of the skeleton builder. As a result, moisture passes through the mud cake easily. Compared to the raw mud, the specific filtration resistance and the humidity of the mud cake are reduced by 97.9% and 19.19%, respectively. Furthermore, Zhou et al.[17] investigated and developed the possibility of dewatering sludge treated with rice husk powder, sodium, and cationic polyacrylamide. Afterward, the results showed sodium chloride dose of 0.25 mol/L, and the dose of rice husk 50% by weight and the humidity of the sludge cake is 29.01%, which increases the dewatering ability of the sludge as a skeleton builder.

Moreover, the presence of a high amount of carbon in the structure of the rice husk indicates the rice's removal organic compounds potential.[18] Pud et al.[13] examined rice husk ash and showed that amorphous silica is its main component, and it is a suitable source of activated carbon. Moreover, Naiya et al.[19] studied the absorption of lead from the activated carbon in rice husk ash and reported the possibility of spontaneous absorption of lead in rice husk ash and the removal efficiency of 96.83 ± 0.30%, which shows the suitability of rice husk ash for wastewater treatment. Furthermore, Moore et al.[20] also showed that rice ash increased the phosphate absorption capacity and the maximum removal (up to 89%) of phosphate from wastewater was at pH = 6 using a dose of 2 g/L for 120 min. Electron microscopy reveals the rough surface of the adsorbent, which may increase the adsorption capacity. The use of rice was also investigated in removing crude oil from wetlands by considering the parameters of particle size, contact surface, and contact time. The results showed that the rate of absorption of crude oil in optimal conditions was about 55%.[21,22] In conclusion, rice husk as a natural material can be widely used in the treatment plant and in the sludge thickening process.

This research was aimed at improving the digestibility of thickening sludge and reducing the costs of the sludge conditioning process. In this research, the mixed ash of rice husk and sludge has been investigated to improve the dewatering and digestibility of sludge in digesters. The main goal of this research was to determine the optimal conditions of using a mixture of rice husk and ash of burnt sludge as a conditioning material from a technical-economic point of view for the thickening and dewatering of biological sludge generated in urban sewage treatment plants using the activated sludge method.

2.2 MATERIALS AND METHODS

2.2.1 QUALITY CHARACTERISTICS OF USED SLUDGE

In the beginning, the biological sludge samples were collected from the return flow line of wastewater treatment plant south of Tehran. The quality characteristics of the biological sludge for the evaluation of the thickening and the dewatering are given in Table 2.1.

TABLE 2.1 Qualitative Characteristics of Biological Sludge of Wastewater Treatment Plant.

Qualitative characteristics	Amount
Ph	6.98 ± 0.2
Chemical oxygen demand (mg/L)	8920 ± 2000
Volatile solids (mg/L)	5400 ± 1000
Volatile suspended solids (mg/L)	4964 ± 1000
Total solids (mg/L)	8270 ± 1000
Total suspended solids (mg/L)	7425 ± 1000
VS/TS	0.75 ± 0.05
Specific resistance filtration (m/kg)	(2.27 ± 0.5) * 10^5
TTF (s)	250 ± 40

To prepare sludge ash by burning it, a dry solids concentration of 20% was obtained from the wastewater treatment plant depot south of Tehran. This sludge has 200 kg dry solids and 800 kg moisture per ton. A blast furnace was used to burn sludge and make ash. Heat is transferred through direct contact with hot gases with wet sludge. As a result, the heat of gases leads the sludge liquid to evaporate and burn the sludge. The furnace was used to burn the sludge at a temperature of 900°C. Finally, the ash from this process was used to condition the sludge.

2.2.2 RICE HUSK PREPARATION

The rice husks were obtained from the agricultural fields of Mazandaran province and processed in modern rice milling factories which consist of two types of low and high yield, as L and H type. The laboratory analysis of both types of husk rice is given in Table 2.2.

2.2.3 CONDITIONING

Rice husk powder:

Two techniques were used to condition the sludge. In the first experiment, rice husk powder was used, and in the second, mud ash and rice husk powder were combined. The pH value of the sludge samples was set in the acidic, alkaline, and neutral ranges, respectively, 4, 10, and 7. This procedure was done to carry out laboratory studies on sludge conditioning by the mechanical method using two types of rice husk. The aim of this process was to investigate the effect of changes in the pH value of the sludge and the dose of rice husk and mud ash on the water separation capability. Then, for each pH value, rice husk powder was injected into the sludge samples with doses of 0.05–0.55 g of dry solids, and the amount mixture of mud ash and rice husk was approximately 0.75–1.25 g. The optimal values of sludge pH and the injection dose of rice husk powder were determined to achieve the minimum amount of filtration time, and capillary suction time of sludge and to compare the effectiveness between the two types of rice husk.

TABLE 2.2 Laboratory Analysis of Two Types of Rice Husks.

Qualitative characteristics	Low yield	High yield
Appearance color	Light brown	Light brown
Physical state	Solid	Solid
Iron (mg/100 g)	157.43	150.61
Humidity (%)	7.25	7.4
Total ash (%)	15.66	19.43
Potassium (mg/100 g)	157.63	112.88
Calcium (mg/100 g)	20	40.1
Magnesium (mg/100 g)	255	230.95
Phosphorus (mg/100 g)	71.61	132.82
Carbohydrate (%)	20.96	35.69
Total fat (%)	0.72	0.4
Protein (%)	5.51	4.28
Nitrogen (%)	0.88	0.68
Crude fiber (%)	49.9	32.8

2.2.4 SPECIFIC RESISTANCE FILTRATION OF SLUDGE

One of the usual methods to determine the ability to separate water from sludge is to test the specific resistance of filtration through the Buchner funnel, because determining the specific resistance filtration of sludge includes a physical procedure like dewatering from pressurized sludge. At each stage of the laboratory tests, the filter paper was slightly moistened to create a watertight status and put it on the Buchner funnel. In the next step, 200 ml of conditioned sludge sample was transferred to the Buchner funnel with a pressure of 50–70 kPa applied by a vacuum pump. In the next step, the time values for the volume of water collected in the bottom container of the Buchner funnel were measured in different volumes (every 25 mL). This process was continued until no more excess water was removed from the sludge and the surface of the sludge cake. Finally, the specific resistance filtration of the sludge is calculated through eq 2.1.[23]

$$SRF = \frac{2P.A^2.b}{\mu.C} \tag{2.1}$$

P: applied vacuum pressure ($N.m^{-2}$)
A: filtration cross-section (m^2)
μ: dynamic viscosity of smoothed sludge ($kg.m^{-1}.s^{-1}$)
C: the condensation of sludge solids *(kg/m³)*
b(t/v^2): sludge rectification time rate

2.2.5 THE SLUDGE CAPILLARY SUCTION TIME (CST TEST)

In this experiment, two circles were drawn with diameters of 15.9 and 22.2 mm on the chromatography paper and placed under a small cylinder with a diameter of 18 mm. The conditioned sludge with a volume of 4.6 mL was poured into the cylinder, and the time for the liquid to reach from the circle with diameter of 15.9 mm to the second circle was measured. Figure 2.1 is shown as a description of the CST test in a pilot form.

2.2.6 SLUDGE VOLUMETRIC INDEX (SVI TEST)

Sludge volumetric index is the amount of volume per milliliter that is occupied by 1 g of a suspension solution after 30 min of settling. The SVI test usually measures the thickening capability of activated sludge and other biological suspensions. For this purpose, it is necessary to determine the thickening of suspended solids in the suspension solution sample (SS) in terms of mg/L and the volume of sludge deposited in the test cylinder after 30 min (SSV) in terms of mL/L. Finally, the volume index of sludge in terms of mL is calculated through eq 2.2.[24]

$$SVI = \frac{SSV \times 1000}{SS} \tag{2.2}$$

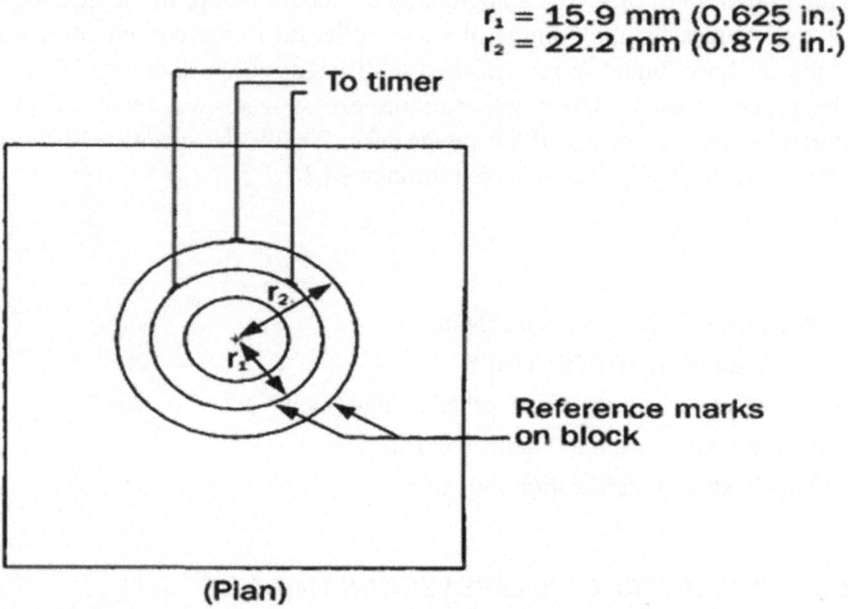

FIGURE 2.1 Laboratory pilot to determine the sludge capillary suction time (CST).

2.2.7 DIGESTIBILITY TEST

First, the samples were conditioned and then centrifuged for 10 min with 4000 rpm. The control sample was also used which was a mixture of 60% primary sludge and 40% secondary sludge. The samples set were placed in the pilot for 7 days and the amount of produced gas, dry solids (DS), and volatile (VS) were measured and evaluated after the digestibility test.

2.2.8 TEST DESIGN MODEL

In this research, the response surface method (RSM) Design-Expert software (Version 12, Inc. Stat-Ease) was used to statistically design the experiment and analyze the data, including the duration of thickening, specific filtration resistance, and capillary suction time. The ultimate goal of the RSM is the simultaneous optimization of the factors affecting the target parameter. This method, considering the number of effective factors and the maximum and minimum limits determined for each factor, forms a matrix of required test situations consisting of the number of tests and levels of each element. The design of the experiment in the response level method was done in a way that even without repeating the experimental conditions, it achieved reliable results. On the other hand, it determines the necessary conditions to provide the best possible answer through the process. The two factors of sludge pH value and conditioning agent injection dose were chosen as the most important influencing factors on the response variables, which are pH in the range of 4–10 and dosage of rice husk powder in the range of 0.05–0.50 g per gram of dry solids (g/gDS). The mixture of sludge ash and rice husk powder was considered in the range of 1.25–0.75 g of dry solids.

The disparities between the observed response values and the expected response values at each combination of factor values are known as residuals. The validity of the model for the currently chosen response can be evaluated using residual plots.[25] In RSM, the user can check the residual plot to show the normality assumption of the models. After studying the relevant equations, the remaining graph was drawn to determine the correctness of the models.

2.3 RESULTS

2.3.1 THE SPECIFIC RESISTANCE FILTRATION (SRF)

In the process of biological sludge conditioning by the mechanical method through rice husk injection, two parameters of sludge pH and rice husk injection dose are considered effective factors on the specific resistance of sludge filtration (SRF). In this regard, the results of laboratory studies on the effect of these two factors, sludge pH and rice husk injection dose, on the filtration resistance of excess biological sludge are shown in Figure 2.2.

For L type rice husk according to Figure 2.2a, when the pH value of the sludge is in the neutral range, by increasing the injection dose of rice husk powder from 0 to 0.05 g, the specific resistance of sludge filtration from 2.27×10^5 decreases to 0.702×10^5 m/kg. The highest amount of decline in the specific resistance filtration for sludge at pH 7 with the injection dose of 0.05 g is equal to 68%. Meanwhile, by increasing the injection dose of rice husk powder in the sludge g/gDS, the specific filtration resistance will have an upward trend, one of the reasons for this phenomenon is the formation of

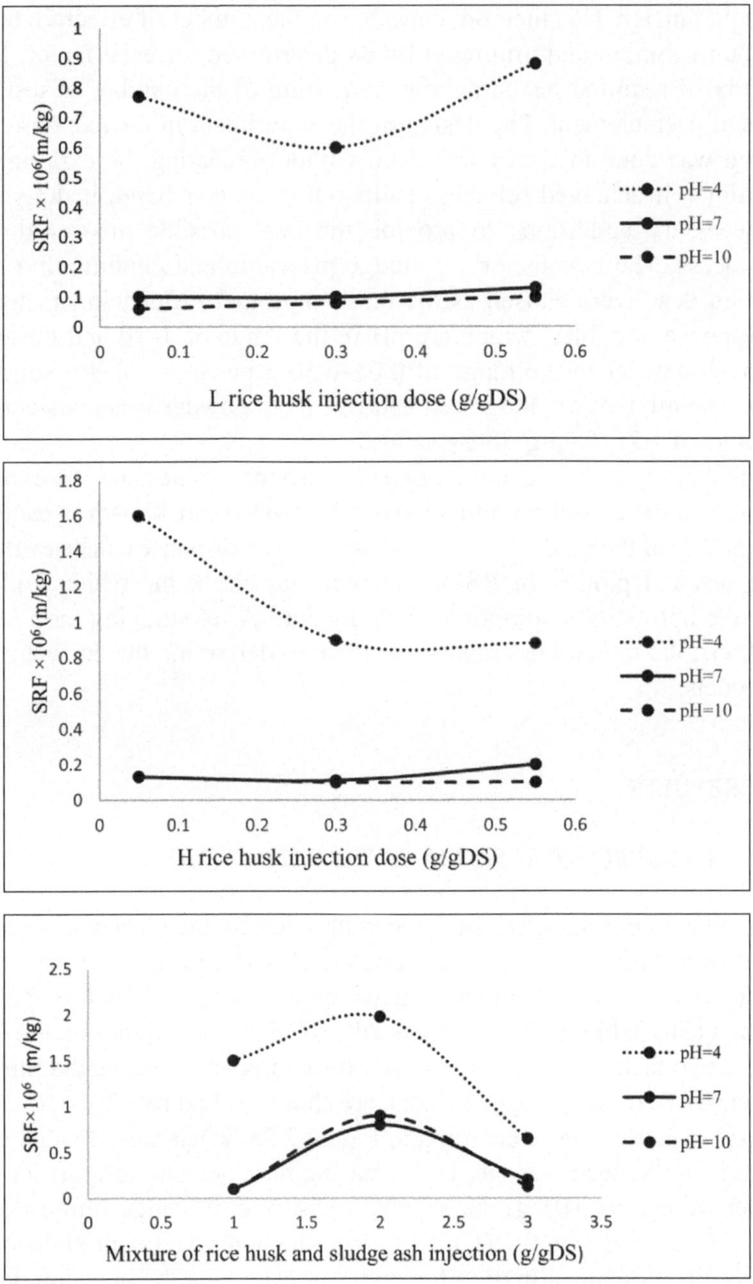

FIGURE 2.2 The influence of the pH value of sludge and the injection dose of (a) L rice husk, (b) H rice husk, and (c) mixture of rice husk and sludge ash on the specific resistance of biological excess sludge filtration. (Dry solids density: 4000 ± 1000 mg/L).

bound water on the surface of free rice husk powder particles in the environment. Free rice husk powder means rice husk particles that are not absorbed on the surface of the sludge and exist freely in the environment. Another reason could be the colloidal behavior caused by rice husk powder particles.[26]

In the acidic pH range, by reducing the pH of the sludge from the neutral range and reaching a pH value of 4, the specific filtration resistance decreases to a small amount, which can be due to the decrease in the negative zeta potential of the sludge particles.[32] Then, by adding the injection dose of L type rice powder to the sludge from 0.05 to 0.3 g/gDS, the SRF decreases and the maximum reduction of SRF is in the injection dose of 0.3 g. As a result, increasing the injection dose of rice husk powder in the sludge led to rising the SRF.

In conditions where the pH value of the sludge is alkaline and equal to 10, the SRF increased extremely, so dewatering was time consuming, which could be due to the strengthening of the negative zeta potential of the sludge particles.[27] Due to its different properties compared to L type of rice husk powder, H type rice husk powder has a different performance in the biological conditioning process.

For H rice husk according to Figure 2.2b, when the pH value of the sludge is in the neutral range and is equal to 7, by increasing the injection dose of rice husk powder from 0 to 0.05 g, the SRF of sludge decreases from 2.27×10^5 to 0.76×10^5 m/kg. Meanwhile, by increasing the injection dose of rice husk powder, the SRF will have an almost linear trend. In the condition that the pH value of the sludge is alkaline and equal to 10, the resistance increased enormously, so that dewatering was hard.

Moreover, the effect of pH value and injection dose of a mixture of sludge ash and rice husk powder on the specific resistance filtration was investigated and it is shown in Figure 2.2c. By increasing the injection dose of rice husk powder from 0.75 to 1 g/gDS when the pH value of the sludge is in the neutral range and equal to 7, it is possible to reduce the specific resistance of sludge filtration from 2.27×10^5 to 0.87×10^5 m/kg. The same injection dose of 1 g resulted in a maximum reduction of 61% in the SRF of sludge at pH 7. The SRF will also increase as the injection dose of rice husk powder in the sludge is increased, possibly for the reasons mentioned above.

2.3.2 THE CAPILLARY SUCTION TIME

In the CST test reducing the duration of sludge filtering means improving the performance of the dewatering process. The results of measuring capillary suction time at different pH values and doses of L and H type of rice husk with sludge ash are reported in Table 2.3.

TABLE 2.3 The Effect of pH Value and Injection Dose of Injection on the Capillary Suction Time (s).

Type of dose	Sludge pH value	Injection dose (g/gDS)		
		0.05	0.3	0.55
L	4	4	4	4
	7	4	3.8	5
	10	7	8	11
H	4	4	5	4.5
	7	5	4	5
	10	7.5	6	9
Rice and sludge ash	4	4.5	4	3.5
	7	4	4.5	4
	10	14.5	7	8.5

The results of the investigations show that the duration of sludge capillary suction under the conditions of pH equal to 7, and the injection dose of L rice husk powder equaling to 0.3 g/gDS have the highest reduction. In such conditions, the duration of sludge capillary suction, which is equal to 6 s in neutral conditions without rice husk powder injection (control sample), is reduced to 3.8 s. In addition to H type, at an acidic pH equal to 4, a decrease in the duration of sludge capillary suction due to lowering the pH has been observed. It is increased by increasing the injection dose from 0.05 to 0.3 g/gDS. By increasing the injection dose to 0.55 g/gDS, it will have an upward trend, which confirms the results of the specific resistance filtration tests and the duration of sludge thickening. The results of the conducted investigations indicate that the duration of sludge capillary suction under acidic conditions and pH equal to 4 and the injection dose of the mixture of sludge ash and rice husk powder is 0.55 g/gDS has the highest reduction.

2.3.3 THE BIOLOGICAL SLUDGE VOLUMETRIC INDEX

One of the standard tests to determine the efficiency of the thickening process is to measure the sludge volumetric index (SVI). In this test, reducing the volume index of sludge indicates improving the performance of the thickening process. The results of the rice husk powder on the ability for biological sludge thickening are presented in Figure 2.3.

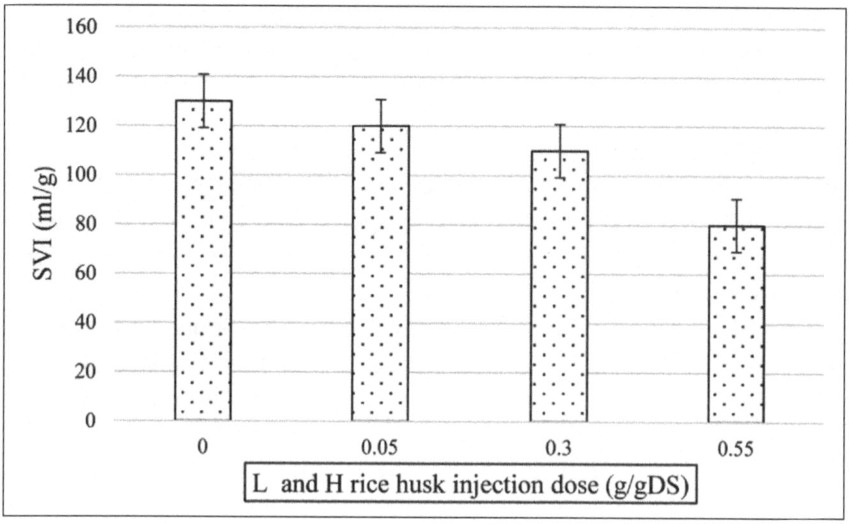

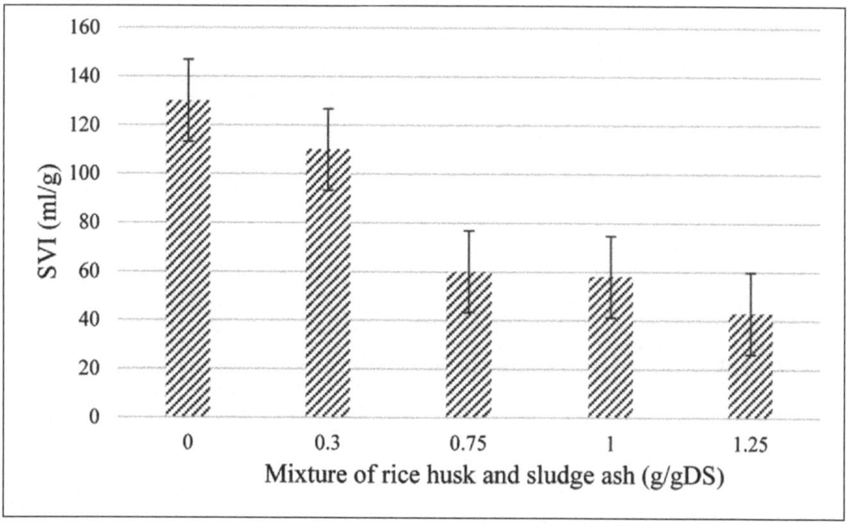

FIGURE 2.3 The effect of the injection dose of (a) rice husk powder and (b) mixture of rice husk and sludge ash on the biological sludge volumetric index.

By increasing the injection dose of rice husk powder into the sludge, its thickening and sedimentation ability is improved. The volumetric index of raw sludge is equal to 130 mg, and as the injection dose of rice husk powder is increased, the volumetric index of sludge declines. The injection dose of 0.55 g/gDS results in the greatest decrease in the volumetric index of sludge. The results of the investigations show that the injection dose of L and H rice husk powder is equivalent to 0.55 g/gDS, indicating the optimal laboratory management conditions. Because increasing the dose of ash injection more than this amount will not have a significant effect on reducing the volume index of sludge.

According to the obtained results, the simultaneous use of acidic pH values and the injection of rice husk powder and sludge ash has a better performance than the injection of rice husk powder alone. In general, it can be said that alkaline environments harm the performance of the sludge dewatering process.

The volumetric index of the raw sludge and rice in Figure 2.3b is 130 mg, and it lowers as the injection dosage is risen. The injection dose of 1.25 g/gDS led to the greatest reduction in sludge volumetric index. Based on the investigation's findings, the ideal laboratory management conditions result in an injection dose of 1 g/gDS of a powdered mixture of rice husk and sludge ash. Since raising the injection dose above this point will not significantly lower the sludge's volume index.

2.3.4 INVESTIGATING THE DIGESTIBILITY OF CONDITIONED SLUDGE

The sludge conditioned with rice husk, biological with a mixture of sludge ash and rice husk were examined to evaluate the efficiency. The results are described in Table 2.4.

TABLE 2.4 Effect of Conditioners on Sludge Digestibility.

Type of conditioning	Before the digestion process		After the digestion process	
	DS (%)	VS (%)	DS (%)	VS (%)
Sludge conditioned with rice husk powder	3.85	87.35	3.81	82.78
Sludge conditioned with a mixture of sludge ash and rice husk powder	3.85	87.35	3.81	64.72

In Table 2.4, the highest reduction in the level of volatile solids in the sludge treated with a mixture is about 22.63%. If this number is close to 50%, it indicates a sludge with greater stabilization and better performance in the digestion process. It is acceptable which has performed well compared to control samples. For the sludge conditioned with rice husk powder, the level of volatile solids is almost equal to the control sample, which indicates that it does not have an appropriate effect on the digestibility process compared to the control samples.

2.3.5 THE RESULTS OF LABORATORY STUDIES USING THE RESPONSE SURFACE METHOD (RSM)

The effective factors on the dewatering ability of sludge along with the levels examined in Design Expert 12 software for used conditioning materials are presented in Tables 2.5 and 2.6.

TABLE 2.5 RSM for L Rice Husk Powder.

Run	Factor 1 A: pH	Factor 2 B: Dose mg/g	Response 1 SRF m/kg	Response 2 CST S
1	7	300	72,703.4	3
2	7	300	71,491.6	5
3	7	300	80,579.6	5
4	4	550	110,267	5
5	10	550	986,040	14
6	7	300	84,214.8	4
7	7	300	86,638.2	4
8	4	300	99,361.3	4
9	10	300	604,347	8
10	7	300	93,302.7	3
11	10	50	689,167	6
12	7	50	70,279.9	4
13	7	300	82,397.2	4
14	10	50	848,206	6
15	4	50	109,661	5
16	10	550	742,180	8
17	7	300	110,873	3
18	4	550	115,720	5
19	7	550	122,990	5
20	4	50	96,937.8	3

TABLE 2.6 RSM for H Rice Husk Powder.

Run	Factor 1	Factor 2	Response 1	Response 2
	A: pH	B: Dose	SRF	CST
		mg/g	m/kg	S
1	10	550	1.0371E + 06	10
2	4	50	102,640	4
3	4	50	89,922.6	4
4	7	300	86,246.2	3
5	4	550	310,059	4
6	7	300	85,533.4	4
7	10	50	1.65069E + 06	8
8	7	300	91,198.1	3
9	4	300	80,953.4	5
10	7	300	103,953	4
11	10	300	902,347	6
12	7	300	86,109.7	4
13	7	50	76,828.4	5
14	10	50	1.52437E + 06	7
15	7	550	80,953.4	5
16	10	550	1.67837E + 06	8
17	4	550	98,851.1	5
18	7	300	77,344	5
19	7	300	60,758	4
20	7	300	75,281.5	4

Among the proposed models of RSM, the quadratic model was chosen as suggested by the software and because it has favorable statistical indicators such as the residual sum of squares. Variance analysis is used to determine the effect of each factor as well as the mutual effect of these factors on the efficiency of the dewatering process.

First, for the L rice husk powder:

$$\text{SRF} = 75369.2 + 333799\ \text{A} + 26294.5\ \text{B} + 21432.4\ \text{AB} + 316108\ \text{A2} + 60889\ \text{B}^2 \quad (2.3)$$

$$CST = 3.8875 + 2\,A + 1.3\,B + AB + 2.0625\,A^2 + 0.5625\,B^2 \tag{2.4}$$

For the H rice husk powder:

$$SRF = 50876.9 + 611044*A + -23912.4*B + -84492.9*AB + 570478*A^2 + 157718*B^2 \tag{2.5}$$

$$CST = 3.9125 + 1.7A + 0.4B + 0.25 * AB + 1.4375 * A^2 + 0.9375 * B^2 \tag{2.6}$$

For the Design Expert software models, three-dimensional diagrams show the simultaneous effect of two parameters on the response variables. By using these diagrams, the simultaneous effect of the response variables can be observed in the states that have not been studied in the laboratory. In the following, the adequacy and statistical accuracy of each of the mathematical models fitted to the results of laboratory studies was evaluated using the statistical charts provided for each of the response variables by the software. According to the residual graph of RSM, the obtained results are approximately 99.3% correct on average.

2.4 RESULTS

1. Low- and high-yield rice husks, as well as a mixture of burned sludge ash and rice husk powder, can be used as biological excess sludge conditioning material for dewatering.
2. The effect of two types of rice husk powder and burnt sludge ash mixture and rice husk powder on the dewatering process from biological sludge and its injection dose is greater in low pH values and acidic environments than neutral and alkaline pH values.
3. The most appropriate injection dose of L rice husk powder in the biological excess sludge conditioning process to achieve optimal laboratory management conditions is in the range of 0.25–0.35 g/gDS.
4. The most appropriate injection dose of H rice husk powder in the biological sludge conditioning process to achieve optimal laboratory management conditions is in the range of 0.2–0.3 g/gDS.
5. The most appropriate injection dose of the mixture of burnt sludge ash and rice husk powder in the biological excess sludge conditioning process to achieve optimal laboratory management conditions is in the range of 0.95–1.05 g/gDS.

KEYWORDS

- **sustainable environment**
- **wastewater treatment**
- **sludge treatment**
- **sludge thickening**
- **dewatering**

REFERENCES

1. Tyagi, V. K.; Lo, S.-L. Sludge: A Waste or Renewable Source for Energy and Resources Recovery?, *Renew. Sustain. Energy Rev.* **2013,** *25*, 708–728.
2. Kazemi, M. J.; Badalians Gholikandi, G. Feasibility Study of Reusing Water Treatment Plant Sludge in Wastewater Treatment Plant Sludge Treatment. *J. Water Wastewater Sci. Eng.* **2022.**
3. Asli, H. H. Investigation of the Factors Affecting Pedestrian Accidents in Urban Roundabouts. *Comput. Res. Prog. Appl. Sci. Eng.* **2022,** *8* (1).
4. Asli, H. H.; Arabani; M. Analysis of Strain and Failure of Asphalt Pavement. *Comput. Res. Prog. Appl. Sci. Eng.* **2022**, *8* (1).
5. Sedaghat, A.; Rabbani; M.; Farrokhi-Asl H. A Sustainable Transportation Location Inventory Routing Problem. *Comput. Res. Prog. Appl. Sci. Eng. (CRPASE)*. **2022,** *8* (3).
6. Abbasi, E.; Pasandideh, H. R.; Zandiyeh, M.; Behboodi, M. In *Optimizing Durability of the High Performance Nano-concrete Applying Taguchi Method and Desirability Function.* Proceedings of the International Conference on Industrial Engineering and Operations Management; Rabat, Morocco, 2017; pp 11–13.
7. Grobelak, A.; Grosser, A.; Kacprzak, M.; Kamizela, T. Sewage Sludge Processing and Management in Small and Medium-sized Municipal Wastewater Treatment Plant-New Technical Solution. *J. Environ. Manag.* **2019,** *234,* 90–96.
8. Qi, Y.; Thapa, K. B.; Hoadley, A. F. Application of Filtration Aids for Improving Sludge Dewatering Properties—A Review. *Chem. Eng. J.* **2011,** *171* (2), 373–384.
9. Wang, C.; Harbottle, D.; Liu, Q.; Xu, Z. Current State of Fine Mineral Tailings Treatment: A Critical Review on Theory and Practice. *Minerals Eng.* **2014,** *58,* 113–131.
10. Lo, I. M.; Lai, K. C.; Chen, G. Salinity Effect on Mechanical Dewatering of Sludge with and without Chemical Conditioning. *Environ. Sci. Technol.* **2001,** *35* (23), 4691–4696.
11. Krüger, O.; Adam, C. Recovery Potential of German Sewage Sludge Ash. *Waste Manag.* **2015,** *45,* 400–406.
12. Sanin, F. D.; Clarkson, W. W.; Vesilind, P. A. *Sludge Engineering: The Treatment and Disposal of Wastewater Sludges*; DEStech Publications, Inc., 2011.
13. Pode, R. Potential Applications of Rice Husk Ash Waste from Rice Husk Biomass Power Plant. *Renew. Sustain. Energy Rev.* **2016,** *53,* 1468–1485.

14. Armesto, L.; Bahillo, A.; Veijonen, K.; Cabanillas, A.; Otero, J. Combustion Behaviour of Rice Husk in a Bubbling Fluidised Bed. *Biomass Bioenergy.* **2002,** *23* (3), 171–179.
15. Wang, S.; Yang, Y.; Chen, X.; Lv, J.; Li, J. Effects of Bamboo Powder and Rice Husk Powder Conditioners on Sludge Dewatering and Filtrate Quality. *Int. Biodeterior. Biodegrad.* **2017,** *124,* 288–296.
16. Wu, Y.; Zhang, P.; Zhang, H.; Zeng, G.; Liu, J.; Ye, J.; Fang, W.; Gou, X. Possibility of Sludge Conditioning and Dewatering with Rice Husk Biochar Modified by Ferric Chloride. *Bioresource Technol.* **2016,** *205,* 258–263.
17. Zhu, C.; Li, F.; Zhang, P.; Ye, J.; Lu, P.; Wang, H. Combined Sludge Conditioning with NaCl-Cationic Polyacrylamide-rice Husk Powders to Improve Sludge Dewaterability. *Powder Technol.* **2018,** *336,* 191–198.
18. Bharadwaj, A.; Wang, Y.; Sridhar, S.; Arunachalam, V. Pyrolysis of Rice Husk. *Curr. Sci.* **2004,** 981–986.
19. Naiya, T. K.; Bhattacharya, A. K.; Mandal, S.; Das, S. K. The Sorption of Lead (II) Ions on Rice Husk Ash. *J. Hazard. Mater.* **2009,** *163* (2–3), 1254–1264.
20. Mor, S.; Chhoden, K.; Ravindra, K. Application of Agro-waste Rice Husk Ash for the Removal of Phosphate from the Wastewater. *J. Cleaner Prod.* **2016,** *129,* 673–680.
21. Thompson, N. E.; Emmanuel, G. C.; Adagadzu, K. J.; Yusuf, N. B. Sorption Studies of Crude Oil on Acetylated Rice Husks. *Arch. Appl. Sci. Res.* **2010,** *2* (5), 142–151.
22. Khodadadi, M. J.; Abbasi, M.; Riahi, S.; Shokrollahzadeh, H. Investigation on Hinetics of Carbon Dioxide Absorption in Aqueous Solutions of Monoethanolamine+ 1, 3-diaminopropane. *Separat. Sci. Technol.* **2019,** *54* (17), 2800–2808.
23. Federation, W. E.; Association, A. *Standard Methods for the Examination of Water and Wastewater*; American Public Health Association (APHA): Washington, DC, USA, 2005, p 21.
24. Daigger, G.; Roper R. Jr. The Relationship Between SVI and Activated Sludge Settling Characteristics. *J. Water Pollut. Control Fed.* **1985,** 859–866.
25. Ahmadi, M. M.; Shahriari, H. Robust Monitoring of Simple Linear Profiles Using M-estimators.
26. Wójcik, M.; Stachowicz, F.; Trzepieciński, T.; Masłoń, A. A.; Opaliński, I. Possibility of Recycling the Biomass Ashes in Sewage Sludge Management. *Arc. Environ. Protect.* **2018,** *44* (3).
27. Gholikandi, G. B.; Zakizadeh, N.; Masihi, H. Application of Peroxymonosulfate-ozone Advanced Oxidation Process for Simultaneous Waste-activated Sludge Stabilization and Dewatering Purposes: A Comparative Study. *J. Environ. Manag.* **2018,** *206,* 523–531.

CHAPTER 3

Application of Biotechnology Tools in Agrarian and Industrial Wastewater Treatment

BHASKAR GAONKAR, ROOPASHREE R., and V. SWARUPA

Department of Chemistry and Biochemistry, School of Sciences, Jain Deemed to be University

ABSTRACT

Water, vital for life despite lacking nutritional value, faces critical management challenges. With only a small fraction available for consumption, agricultural reliance exacerbates scarcity. Industrial pollutants worsen water quality, threatening ecosystems and human health globally. Untreated wastewater spreads disease, depletes oxygen, and contaminates with heavy metals and chemicals, jeopardizing aquatic life. Advanced biotechnological tools offer sustainable solutions, mitigating traditional drawbacks like sludge formation. Phytoremediation, bioreactors, and membrane bioreactors emerge as efficient treatments. As populations grow, effective wastewater management becomes crucial for sustaining water resources amid urbanization, necessitating holistic and technologically advanced approaches.

3.1 INTRODUCTION

All the living organisms and hydrosphere of the earth are predominantly made up of water. Interesting fact is that, even though it does not have any

Sustainable Water Engineering: Smart and Emerging Technologies. Ali Pourhashemi, Kian Hariri Asli, Ann Rose Abraham, & A. K. Haghi (Eds.)

calorific significance or nutritive value, it is vital for all the living systems on earth.[1,2] The statistics says, approximately 97% of the water present on earth is salt water (in ocean) and 2% as ice. So, there is only 1% water available for human and animal consumption. Water shares an intensely interconnected network (like food chain) with earth and air. If we spray something into the air or pour something on ground it finally enters water bodies. The water quality is one of the defining factors of socioeconomic advancement, healthy ecosystem, and health of the individuals living in that environment.[1–3]

Agriculture sector itself accounts for 92% of water usage. Rivers and underground water provide 70% of the water used for irrigation. According to a recent survey, 40% of the world inhabitants are situated in substantial water-scarce zones, indicating the water crisis for irrigation. Hence, recycling wastewater for irrigation is an ideal way of substituting for freshwater consumption in agriculture.[4,5]

It is a fact that global economy is immensely influenced by the industrial revolution, but it eventually became burden to the ecosystem. The shift from farming to industrialization posed huge damage to the environment and ecosystems, which were worsened by the growth of industries. The refractory pollutants expelled from agrarian and industrial wastewater substantially disrupted the ecosystems.[6–8]

The mother nature can efficaciously handle minute quantity of toxins or pollutants released into the environment. But in reality, billions to trillion gallons of sewage water has laden on to the nature without being treated. The right way is to process the wastewater to cut down the pollutants and toxins to a level nature can handle. The untreated or poorly treated wastewater can be a major threat to the stability of the ecosystem and ultimately to the mankind. When we look at the global statistics, around 80% of the sewage water reach water bodies untreated or without being reused because of which nearly 1.8 billion individuals are suffering from various life-threatening ailments.[4–5]

The deleterious effect of untreated water on ecosystem is countless. Suspended particles and putrefying organic matter in the wastewater can eat up dissolved oxygen present in the water, major cause for the death of the aquatic creatures. The high influx of nutrients particularly nitrogen and phosphorus may create hypertrophic conditions or over-fertilization of water bodies, leading to excessive plant and algal growth, lower oxygen availability to hypoxic levels (in the night), destroy breeding grounds, and change habitat leading to a decline in several aquatic dwellers. There may be acute or chronic side effect of heavy metals such as arsenic, chromium, mercury, lead, and cadmium on different aquatic animals. The inorganic chloramines and organic chlorine compounds, pest controlling agents, and fertilizers expelled from industrial wastes and agricultural practices can be toxic to aquatic population.[4–8]

With expanding population, the natural habitats are being damaged. The wastewater management strategy holds prime importance as it is intertwined with accessibility to safe and adequate water supplies. Upgrading the way we treat wastewater is also crucial in his regard. Agriculture, industrial development, energy generation, and rapidly expanding cities cause acute water paucity which can be tackled with proper planned wastewater recycling techniques.[9,10]

Wastewater can be handled using various physical and chemical approaches. The major shortcomings of these methods comprise large quantity of sludge formation, need of chemical agents, overpriced, etc. These drawbacks were managed with the discovery of advanced biotechnical tools that became routine treatment strategies to handle wastewater. Phytoremediation, bioaugmentation, bioremediation, and bioreactors are some of the major biological techniques available which work under aerobic and/or anaerobic environments. The techniques that are substantially constructed on biotechnology and permit effective wastewater treatment with minimum ecological hazard include anaerobic treatment, rotating biological contactors, membrane bioreactor, activated sludge, biocatalyst, membrane biofilm reactor, and trickling filters. The quantity and quality of the wastewater to be treated is the predominant criteria in selecting traditional and nonconventional treatment strategies.

3.2 ACTIVATED SLUDGE

Edward Arden and W.T. Lockett from United Kingdom developed an activated sludge method to treat domestic wastewater in the year 1913. At present this method is used to treat wastes from communal, petroleum, paper, food, and chemical industries. The wastewater is treated with aerobic microorganisms and oxygen supply.[11–14] The growth and multiplication of microorganisms are facilitated by the carbon source and energy present in the organic contaminants. Nitrogen and phosphorus are added as external supplements to assist the growth of microorganisms. The major organic nutrients are eliminated by aerobic bacteria (95%) and the remaining trace (5%) by protozoa, metazoa, filamentous bacteria, algae, and fungi.[14–17]

3.2.1 CONSTITUENTS OF ACTIVATED SLUDGE

The three basic constituents of activated sludge are listed below:

1. Aeration tank: It serves as a bioreactor and aids in the uniform mixing of activated sludge-wastewater.
2. Settling tank: It is constructed to provide roughly around 2–4 h of hydraulic detention time. It acts as a barrier between the activated sludge and treated wastewater.
3. Return activated sludge: Ample quantity of microorganisms in the aeration tank is maintained by return activated sludge where it transfers the sludge from the clarifier to aeration basin.

3.2.2 WORKING PRINCIPLE OF ACTIVATED SLUDGE PROCESS

Activated sludge method is the general method to treat sewerage. In this practice wastewater and microorganisms (mixed liquor) are fused with air or oxygen in which the solid particles are broken and released into an activated sludge chamber. The growth and multiplication of microorganisms depend on the organic nutrients present in the wastewater. The bacteria secret a sticky slime layer around the cell wall resulting in the formation of cluster called as bio-solids/sludge/floc which separate from the liquid phase. The floc formed make use of organic content present in the wastewater. The bio-solids formed by the bacteria settle to the bottom of the settling basin. The process is repeated as some portion of the sludge is again circulated from the clarifier to aeration basin to ensure sufficient microbial population (Figure 3.1). This method removes about 85–95% of solid wastes. Climatic factor and chemical composition present in sewerage influence the effectiveness of this process. Industrial wastewater must be pretreated to ensure the removal of toxic chemicals before subjecting into this process. Soaps and detergents can develop froth and cause trouble. Yet activated sludge is the productive and general processes for the treatment of sewerage.[11–17]

3.2.3 THE PROS OF ACTIVATED SLUDGE TREATMENT

- Equipment setup cost is low.
- The water obtained from this method is of fine quality if the wastewater is of invariable type.

- Meager space for installation.
- Exempt from flies and odor nuisance.

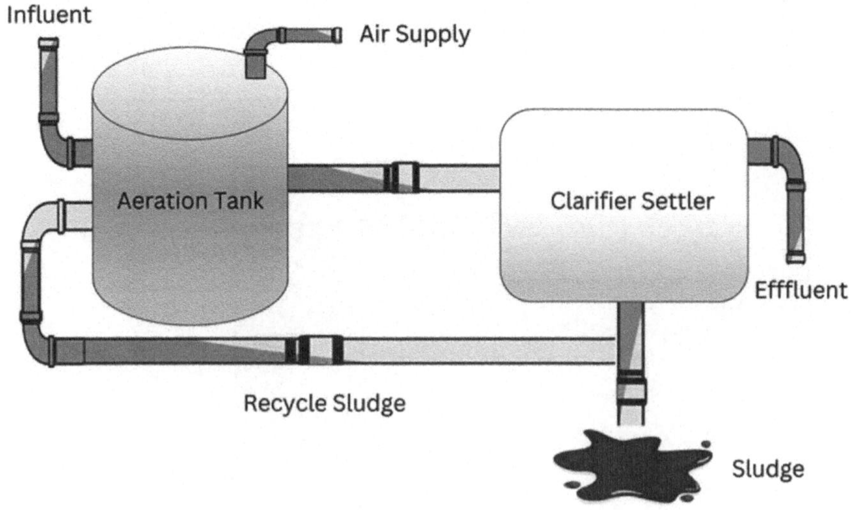

FIGURE 3.1 Activated sludge.

3.2.4 THE CONS OF ACTIVATED SLUDGE TREATMENT

- This method is sensitive as it depends on the chemical composition of wastewater.
- Challenge in sludge disposal.
- Requires skilled designer.

3.3 MEDIA FILTERS (TRICKLING FILTERS)

Trickling filter method is one of the long-established methods used to get rid of soluble organic constituents present in wastewater by employing microorganisms.[16,18–20] These microbes develop a biological film or slime layer measuring around 0.1–0.2 mm thickness. Trickling filter method is termed as attached-growth processes/biofilm reactors/biofilters/percolating/sprinkling filter. This is a secondary treatment process as it is associated with other techniques like rotating biological contactors and packed bed reactors (bio-towers) (Figure 3.2).

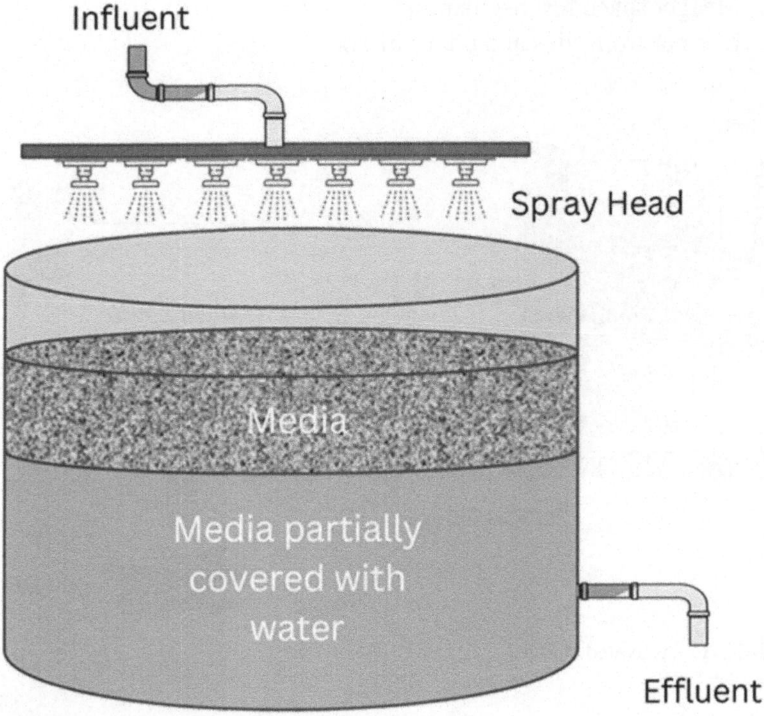

FIGURE 3.2 Trickling filter.

3.3.1 CLASSIFICATION OF TRICKLING FILTERS

It is classified as high-rate and low-rate. The classification is mainly centered on hydro mechanical or organic loading.[18–22]

3.3.1.1 HIGH-RATE FILTERS

The filter bed is loaded with secondary treated wastewater where hydro mechanical and organic loads are applied at the higher rate. Large volume of waste per acre of surface area of the filter assists the BOD of organic content. The recirculation of effluent back to filter accounts for the effectiveness of this method as it increases the exposure time with the microorganisms. Recirculation can either be carried out continuously or fragmentary. The advantage of recirculation is that it decreases the difficulty associated with ponding and constraint of ventilation.

3.3.1.2 LOW-RATE FILTERS

The other names are conventional trickling filter/ordinary trickling filter/ standard rate filter. Rotatory distributors are amalgamated with these filters which are circular. Rectangular filters are gaining importance in the present day. The grade of the effluvium generated from this method is of varied influent strength. The biological oxygen demand of the waste water put into the filter depends on the organic load. In this process the stabilized solids are removed from the secondary treated wastewater. This water passes through the filter bed which filters the water, collected in the sedimentation tank.

3.3.2 WORKING PRINCIPLE OF TRICKLING FILTERS

Filter bed is basically made up of rocks or gravel. Slag, polyurethane foam, sand, sphagnum peat moss, or plastic media can also be used as filter bed. The settled waste effluent is obtained from primary sedimentation tank. This wastewater is passed through the filter bed. The rate of growth of microorganisms and the film formation depends on the diffusion rate of the wastewater which involves both absorption and adsorption of the organic wastes. Diffusion or splashing favors the aerobic condition required for the growth of the microorganism. Natural convection can also be applied provided that the filter medium is porous.[18] The microbial slime produces oxidized products, carbon dioxide, and water by breaking down organic substrates in the presence of oxygen. Sloughing is a process where the microbial layer gets detached from the filter due to the thickening of the microbial layer. As the microbial layer thickens the inner layer of microbial colony becomes anaerobic due to the lack of oxygen. Slough formation depends on the formation of slime layer by microbial activity. Clarifier or sedimentation tank aids in the removal of formed slough.[19–22]

3.3.3 THE PROS OF TRICKLING FILTER

- This method requires minimum land with high rate of filter load.
- Media requirement is less.
- The quality of water obtained after purification is good.
- Efficient and easy to operate.
- No requirement of expert supervision.
- Self-cleaning and easy operation.
- Requirement of mechanical equipment is less.

3.3.4 THE CONS OF TRICKLING FILTER

- Construction budget is high.
- These filters require primary sedimentation; hence, raw sewerage cannot be treated directly.
- Breeding of nuisance flies.
- Prevail of awful odor.

3.4 BIOCATALYST IN THE TREATMENT OF WASTEWATER

Enzymes are proteins produced in the living cells which are biocatalyst for a specific biochemical reaction. The application of enzymes was first put forward in the year 1930 for sustainable treatment of wastewater.[4,23–26]

As these enzymes are highly specific and selective, they are employed to treat both organic and inorganic contaminants present in the wastewater (Table 3.1). Trace amounts of contaminants can also be treated with this method. A specific concentration of enzymes is introduced directly or in combination with the microbes. These enzymes can also be obtained from the plant source. The enzymes have the potential to convert the toxic substrates into harmless or less toxic substances.[26–29]

TABLE 3.1 Microbial Enzymes and Pollutants.

Enzymes	Pollutants
Oxygenase	Petrol and diesel
Peroxidase	Phenols, dyes, aromatic compounds, and organic peroxides
Polyphenol oxidases	Phenols
Tyrosinase	Phenols
Laccases	Dyes and phenol
Cyanidase and cyanide hydratase	Cyanide
Chloro-peroxidase	Phenols
Lignin peroxidase	Phenols and aromatic compounds
Parathion hydrolase	organophosphate pesticides
Mn-peroxidase	monoaromatic phenols and aromatic dyes
Phosphatase	Heavy metals
Lactases	Dairy wastes
Alkylsulfatase	Surfactant

3.4.1 THE PROS OF BIOCATALYST IN WASTEWATER TREATMENT

- Easy to handle.
- Specific for a particular substrate.
- Effective for the removal of contaminants present in trace amount.
- Reduces or eliminates the toxic chemicals present in wastewater.
- Leads to sustainability as they are biodegradable.
- Green method.

3.4.2 THE CONS OF BIOCATALYST IN WASTEWATER TREATMENT

- Sludge disposal.
- May get inhibited by inhibitors present in the wastewater.

3.5 ANAEROBIC WASTEWATER TREATMENT

This is a process in which microorganisms are employed to digest organic pollutants in the absence of oxygen. In the treatment cycle the bioreactor that receives wastewater contains sludge (a semi-solid, soft, and thick material), which harbor some of the anerobic bacteria and other beneficial microbes that act on the biodegradable materials present in the wastewater.[8,18] This process drops the total suspended solids, biogas byproducts, chemical oxygen demand, and biological oxygen demand of the effluent formed (Figure 3.3).

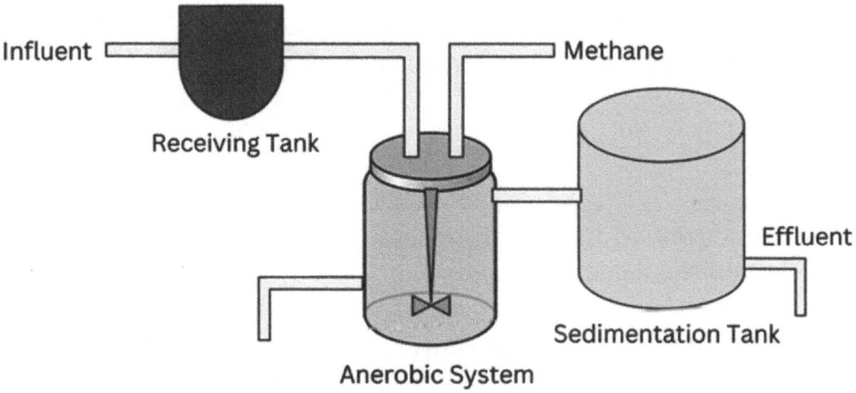

FIGURE 3.3 Anaerobic wastewater treatment.

3.5.1 STAGES OF THE PROCESS

There are two stages in the anaerobic treatment.

- Acidification phase or acid-forming phase.
- Methane production phase.

3.5.1.1 ACIDIFICATION PHASE

It is the early stage of the anaerobic wastewater treatment in which the complex organic matter is completely broken doswn to produce simpler volatile molecules.

3.5.1.2 METHANE PRODUCTION PHASE

This stage mainly consists of two steps:

The first step is acetogenesis phase, in which anerobic organisms act on the sludge to form carbon dioxide, hydrogen gas, and acetate. The second step is methanogenesis where the products of acetogenesis are further metabolized to form carbon dioxide and methane. The wastewater is further processed or expelled out of the system, while the gas (methane) can be separated and used as a fuel. The anaerobic digester systems can be designed to have acidification tank and bioreactor in a single or separate unit based on the specific requirement.[30,31]

3.5.2 CONVENTIONAL ANAEROBIC WASTEWATER MANAGEMENT SYSTEMS CONSIST OF THE FOLLOWING TYPES

3.5.2.1 ANAEROBIC LAGOONS

These are huge artificial ponds up to 20 feet depth, ranging between 1 and 2 acres. The wastewater is released into these lagoons where the sludge settles down leaving the liquid layer on the top. The microorganisms degrade the organic contaminants present in the sludge under anoxic conditions. This is a primary treatment strategy for the agricultural, industrial, and municipal wastewater.[30–33]

3.5.2.2 ANAEROBIC SLUDGE BLANKET REACTORS

In this system the sewerage water is allowed to move across a free-floating blanket of suspended sludge particles. The anaerobic organisms in the suspended sludge multiply using the organic contaminants, grow as big granules, and settle at the base of the reactor which can be recycled. The treated effluent flows out of the reactor which may be further processed or expelled out. This process produces biogas which can be collected using collection hoods and used as an alternate energy source.[33–34]

3.5.2.3 ANAEROBIC FILTER REACTORS

This system is composed of a reactor chamber fitted with a fixed filter medium, made up of bricks, gravel, plastic films and particles, and other materials (based on the system type). This filter media is inoculated with microbes which form a biofilm over the filter media. This biofilm may take several months to establish itself to become a fully functional treatment system. The filter media captures particles from the wastewater stream. The anaerobes get ample surface area to make contact and digest the organic material present in the wastewater.[30–34]

3.5.3 PROS OF ANAEROBIC WASTEWATER TREATMENT

- The anaerobes convert organic contaminants into valuable biogas. In some cases, it can completely cater the energy requirement to run the water purification system.
- Compared to the aerobic wastewater purification system the sludge growth in this system is 4–5 times lower.
- The colonies of anaerobes hibernate (without significant mortality) if the sludge is not fed for a long time. The system resumes quickly after the influx of a new batch of wastewater.
- Compared to the aerobic wastewater purification system the volumetric load of this system is 5–10 times higher.

3.5.4 CONS OF ANAEROBIC WASTEWATER TREATMENT

- Less robust system with regards to toxicity and inhibition.

- Incomplete breakdown of organic compounds: The effluent produced in this process may require additional treatment step (like aerobic treatment) to achieve significant decontamination.
- The biogas produced in the system poses odor problems.
- Heating of the influent is essential in most cases, as the process is highly effective between 30 and 37°C.

3.6 MEMBRANE BIOREACTOR

Membrane bioreactor is a multi-stage unit consisting of a bioreactor coupled with a separating membrane filter. Each of these components serves specific purpose.[8] The bioreactor carries out digestion of organic pollutants using microorganisms.[31] But the membrane module filters out the microorganisms and suspended particles to form an effluent. The activated sludge settling property is the major determinant of final clarification step in the conventional treatment plants.[32–34] However, the membrane bioreactor stands out among all the treatment strategies as the membranes serve as a physical barrier and sieve all the suspended solid particles from the effluent. This creates a permeate free from all the suspended matter and microbes. This promotes recycling activated sludge.[35–41] The membrane system is thus an additional advantage and a major difference between the conventional and modern treatment techniques (Figure 3.4)

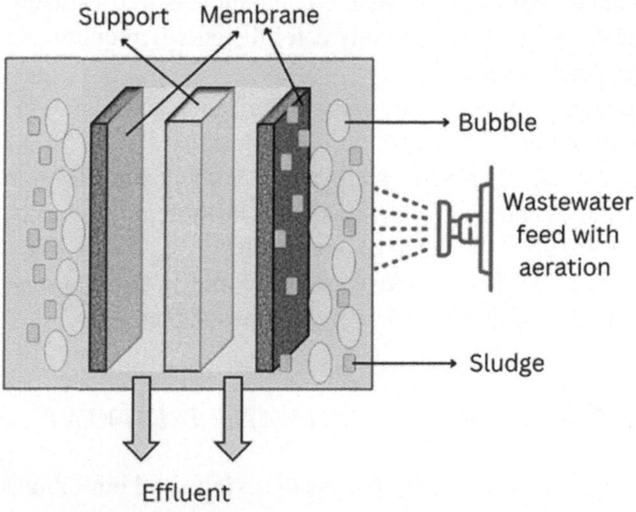

FIGURE 3.4 Membrane bioreactor.

3.6.1 PROS OF MEMBRANE BIOREACTOR

- Withstands extreme temperature and other environmental factors like chemicals and pH.
- Activation of sludge is much easier and efficient compared to traditional systems.
- Inert to microbial degradation.
- It is easy to clean the unit after fouling.

3.6.2 CONS OF MEMBRANE BIOREACTOR

- Construction of the unit is expensive compared to traditional systems.
- The shear-related effects may deactivate the enzymes.
- The microbial enzymes may be inhibited by the products produced in the reaction.
- The enzymatic activity may also be reduced due to the loss of activators and/or cofactors during filtration step.

3.7 ROTATING BIOLOGICAL CONTACTORS

This is an alternative to activated sludge technique and known as biofilm rector, media reactor, surface reactor etc. The setup is typically composed of several discs mounted on a horizontal rod. Several such units can be assembled in series or parallel. The surface of the discs is coated with media, which serves several functions (It is primarily involved in the aeration of wastewater and offers a bed for the growth of the microbial population). With the help of a motor, along with the compressed air the setup is made to rotate at a speed of 1–10 rpm, which creates a thin film of liquid on the surface of each disc.[36] A part of this setup is exposed to air which helps in aeration of liquid film and permits digestion of contaminants by aerobic microbes inoculated on the surface of the disc.[42–44] Thus, the wastewater gets circulated between the liquid and the gaseous phase. The excess biomass is shed from the media when the disc is passed through the liquid waste with the shear stress exerted on the biomass. This enables a balanced microbial growth on the media and prevents clogging.[45–46]

3.7.1 PROS OF ROTATING BIOLOGICAL CONTACTORS

- Low sludge production.
- Brief retention time.
- Construction of the unit is inexpensive.
- Work well with varied range of liquid waste.
- Highly controlled method.
- As the active surface is large, only small contact periods will be sufficient.
- The sluffed biomass can be easily separated from the effluent.

3.7.2 CONS OF ROTATING BIOLOGICAL CONTACTORS

- Absence of provisions for recirculation.
- Low dissolved oxygen transfer efficiency.
- The biomass is sensitive to industrial wastes.
- Mechanical drive units require frequent maintenance.
- Covering the unit is necessary to prevent freezing in extremely cold climates.

3.8 CONCLUSION

The key idea of wastewater treatment is to convert the domestic and industrial effluvium into harmless or less toxic without affecting the environment or living beings. The biotechnological tools can be used to achieve desired quality of water which in turn helps in the reuse of water for either domestic or industrial purpose. The desired quality of water can be achieved by making use of sophisticated instruments and technical hand with high operating costs and being expensive to run. The challenge is to handle the wastewater with low cost, productive, and eco-friendly manner.

Green technology is the primary step toward sustainable approach of wastewater treatment. This method holds good to areas with low- and high-density populations. The first action toward sustainability must be to lower the synthetic and destructive human activities on the environment using eco-friendly treatment strategies. Biotechnology, a science integrates several fields of science from biochemistry to engineering, is a promising way to achieve sustainability. It uses immense capacities hidden in cultured tissue cells and microorganisms to attain sustainable wastewater treatment. Biotechnical approaches replaced or enhanced the conventional approaches (of wastewater treatment) with the innovative systems, methods, and purposes of wastewater treatment. These alternative eco-friendly technological advances allowed less human impact on the world, with reduced ecological imbalance. This has also reduced expenses of handling wastewater and pollution due to effluent release.

One must develop the mindset to look wastewater as a valuable source than a futile matter. Apart from a potential source of usable water, the sewage water can also be a fruitful source of green energy, valuable nutrients, and other recoverable materials. There are several biotechnological tools for wastewater treatment, but more innovations and discoveries are the need of the hour to decrease the impact of human on environment to a near zero level.

KEYWORDS

- **water management**
- **wastewater recycling**
- **urbanization challenges**
- **ecosystems**
- **industrial pollution**
- **sustainable solutions**
- **global water scarcity**

REFERENCES

1. Clemens, M.; Khurelbaatar, G.; Merz, R.; Siebert, C.; van Afferden, M.; Rödiger, T. Groundwater Protection under Water Scarcity; from Regional Risk Assessment to Local Wastewater Treatment Solutions in Jordan. *Sci. Total Environ.* **2020,** *706,* 136066.
2. Hoekstra, A. Y.; Mekonnen, M. M. The Water Footprint of Humanity. *Proc. Natl. Acad. Sci.* **2012,** *109* (9), 3232–3237.
3. Tanji, K. K.; Kielen N. C. Agricultural Drainage Water Management in Arid and Semiarid Areas. Irrigation and Drainage Paper 61. Food and Agriculture Organization of the United Nations, Rome, Italy. 2002.
4. Pedrero, F.; Kalavrouziotis, I.; Alarcón, J. J.; Koukoulakis, P.; Asano, T. Use of Treated Municipal Wastewater in Irrigated Agriculture—Review of Some Practices in Spain and Greece. *Agric. Water Manag.* **2010,** *97* (9), 1233–1241.

5. Tao, C. W., Hsu, B. M., Ji, W. T., Hsu, T. K., Kao, P. M., Hsu, C. P.; Huang, Y. L. Evaluation of Five Antibiotic Resistance Genes in Wastewater Treatment Systems of Swine Farms by Real-time PCR. *Sci. Total Environ.* **2014,** *496,* 116–121.
6. Contreras, J. D.; Meza, R.; Siebe, C.; Rodríguez-Dozal, S.; López-Vidal, Y. A.; Castillo-Rojas, G.; ... Eisenberg, J. N. Health Risks from Exposure to Untreated Wastewater Used for Irrigation in the Mezquital Valley, Mexico: A 25-year Update. *Water Res.* **2017,** *123,* 834–850.
7. Rathna, R.; Varjani, S.; Nakkeeran, E. Sequestration of Heavy Metals from Industrial Wastewater Using Composite ion Exchangers. In *Applications of Ion Exchange Materials in the Environment*; Springer: Cham, 2019; pp 187–204.
8. Hu, Y.; Cheng, H.; Ji, J.; Li, Y. Y. A Review of Anaerobic Membrane Bioreactors for Municipal Wastewater Treatment with a Focus on Multicomponent Biogas and Membrane Fouling Control. *Environ. Sci: Water Res. Technol.* **2020,** *6* (10), 2641–2663.
9. Dutta, A.; Davies, C.; Ikumi, D. S. Performance of Upflow Anaerobic Sludge Blanket (UASB) Reactor and Other Anaerobic Reactor Configurations for Wastewater Treatment: A Comparative Review and Critical Updates. *J. Water Suppl: Res. Technol.-Aqua.* **2018,** *67* (8), 858–884.
10. Mucha, Z.; Wójcik, W.; Polus, M. *Brief Review of Operation of Anaerobic Wastewater Treatment with Membrane Bioreactors*, In E3S Web of Conferences; Vol. 86; EDP Sciences, 2019; p 00020.
11. Gallucci, F.; Basile, A.; Hai, F. I. Introduction-A Review of Membrane Reactors. 2011.
12. Patwardhan, A. W. Rotating Biological Contactors: *Rev. Indus. Eng. Chem. Res.* **2003,** *42* (10), 2035–2051.
13. Gullicks, H.; Hasan, H.; Das, D.; Moretti, C.; Hung, Y. T. Biofilm Fixed Film Systems. *Water.* **2011,** *3* (3), 843–868.
14. Christenson, L.; Sims, R. Production and Harvesting of Microalgae for Wastewater Treatment, Biofuels, and Bioproducts. *Biotechnol. Adva.* **2011,** *29* (6), 686–702.
15. Ezeonu, C. S.; Tagbo, R.; Anike, E. N.; Oje, O. A.; Onwurah, I. N. Biotechnological Tools for Environmental Sustainability: Prospects and Challenges for Environments in Nigeria—A Standard Review. *Biotechnol. Res. Int.* **2012,** *2012.*
16. Shchegolkova, N. M.; Krasnov, G. S.; Belova, A. A.; Dmitriev, A. A.; Kharitonov, S. L.; Klimina, K. M.; Kudryavtseva, A. V. Microbial Community Structure of Activated Sludge in Treatment Plants with Different Wastewater Compositions. *Front. Microbiol.* **2016,** *7,* 90.
17. Nielsen, P. H.; Thomsen, T. R.; Nielsen, J. L. Bacterial Composition of Activated Sludge-importance for Floc and Sludge Properties. *Water Sci. Technol.* **2004,** *49* (10), 51–58.
18. Sustarsic, M. Wastewater Treatment: Understanding the Activated Sludge Process. In *Safety in Ammonia Plants and Related facilities Symposium*; 2009; pp 26–29.
19. Darweesh, M.; Zedan, A. M.; El-Banna, A.; Elbasiuny, H.; Elbehiry, F. Biotechnology for Green Future of Wastewater Treatment. 2021.
20. Buchanan, J. R. Decentralized Wastewater Treatment. 2014.
21. Vallero, D. A. *Applied Ecology. Environmental Biotechnology*, 2nd ed.; 2016; pp 293–358.
22. Pandey, K.; Singh, B.; Pandey, A. K.; Badruddin, I. J.; Pandey, S.; Mishra, V. K.; Jain, P. A. Application of Microbial Enzymes in Industrial Wastewater Treatment. *Int. J. Curr. Microbiol. Appl. Sci.* **2017,** *6* (8), 1243–1254.
23. Zdarta, J.; Jesionowski, T.; Pinelo, M.; Meyer, A. S.; Iqbal, H. M.; Bilal, M.; ... Nghiem, L. D. Free and Immobilized Biocatalysts for Removing Micropollutants from Water and Wastewater: Recent Progress and Challenges. *Bioresource Technol.* **2022,** *344,* 126201.

24. Vayenas, D. V. Attached Growth Biological Systems in the Treatment of Potable Water and Wastewater. *Comp. Biotechnol.* **2011,** *6* (2), 371–383.
25. Cheremisinoff, N. P. Biotechnology for Industrial and Municipal Wastes. *Biotechnol. Waste Wastewater Treat.* **1996.**
26. Buyukgungor, H.; Gurel, L. The Role of Biotechnology on the Treatment of Wastes. *Afr. J. Biotechnol.* **2009,** *8* (25).
27. Sadoff, C.; Muller, M. Water Management, Water Security and Climate Change Adaptation: Early Impacts and Essential Responses. Stockholm: Global Water Partnership; 2009.
28. Aitken, M. D. Waste Treatment Applications of Enzymes: Opportunities and Obstacles. *Chem. Eng. J.* **1993,** *52* (2), B49–B58.
29. Oakley, S.; von Sperling, M. Media Filters: Trickling Filters and Anaerobic Filters. Global Water Pathogens Project; Part, 4; 2017.
30. Beya, C. L.; Kanwugu, O. N.; Ivantsova, M. N. Modern Biotechnological Methods in Wastewater Treatment: A Review. *Chimica Techno. Acta.* **2022,** *9,* Supplement, 9 (Supplement).
31. Rao, M. A.; Scelza, R.; Scotti, R.; Gianfreda, L. Role of Enzymes in the Remediation of Polluted Environments. *J. Soil Sci. Plant Nutr.* **2010,** *10* (3), 333–353.
32. Feng, S.; Ngo, H. H.; Guo, W.; Chang, S. W.; Nguyen, D. D.; Cheng, D.; ... Liu, Y. Roles and Applications of Enzymes for Resistant Pollutants Removal in Wastewater Treatment. *Bioresource Technol.* **2021,** *335*, 125278.
33. Pandey, K.; Singh, B.; Pandey, A. K.; Badruddin, I. J.; Pandey, S.; Mishra, V. K.; Jain, P. A. Application of Microbial Enzymes in Industrial Wastewater Treatment. *Int. J. Curr. Microbiol. Appl. Sci.* **2017,** *6* (8), 1243–1254.
34. Yu, G. H.; He, P. J.; Shao, L. M.; Lee, D. J. Enzyme Activities in Activated Sludge Flocs. *Appl. Microbiol. Biotechnol.* **2007,** *77* (3), 605–612.
35. Bressani-Ribeiro, T.; Almeida, P. G. S.; Volcke, E. I. P.; Chernicharo, C. A. L. Trickling Filters Following Anaerobic Sewage Treatment: State of the Art and Perspectives. *Environ. Sci: Water Res. Technol.* **2018,** *4* (11), 1721–1738.
36. Einschlag, F. S. G.; Carlos, L.; Eds. Wastewater: Treatment Technologies and Recent Analytical Developments. *BoD–Books on Demand*; 2013.
37. Goswami, L.; Kumar, R. V.; Borah, S. N.; Manikandan, N. A.; Pakshirajan, K.; Pugazhenthi, G. Membrane Bioreactor and Integrated Membrane Bioreactor Systems for Micropollutant Removal from Wastewater: A Review. *J. Water Proc. Eng.* **2018,** *26,* 314–328.
38. Marrot, B.; Barrios-Martinez, A.; Moulin, P.; Roche, N. Industrial Wastewater Treatment in a Membrane Bioreactor: A Review. *Environ. Prog.* **2004,** *23* (1), 59–68.
39. Radjenović, J.; Matošić, M.; Mijatović, I.; Petrović, M.; Barceló, D. Membrane Bioreactor (MBR) as an Advanced Wastewater Treatment Technology. *Emer. Contamin. Indus. Municipal Waste.* **2007,** 37–101.
40. Xiao, K.; Liang, S.; Wang, X.; Chen, C.; Huang, X. Current State and Challenges of Full-scale Membrane Bioreactor Applications: A Critical Review. *Bioresour. Technol.* **2019,** *271,* 473–481.
41. Ngo, H. H.; Guo, W.; Surampalli, R. Y.; Zhang, T. C.; Eds. *Green Technologies for Sustainable Water Management*; American Society of Civil Engineers; VR, Reston, VA, USA, 2016.
42. Tay, J. H.; Liu, Y.; Tay, S. T. L.; Hung, Y. T. Aerobic Granulation Technology. *Adv. Biol. Treat. Proc.* **2009,** 109–128.

43. Pal, S.; Sarkar, U.; Dasgupta, D. Dynamic Simulation of Secondary Treatment Processes Using Trickling Filters in a Sewage Treatment Works in Howrah, West Bengal, India. *Desalination.* **2010,** *253* (1–3), 135–140.
44. Singh, R. L.; Ed.. *Principles and Applications of Environmental Biotechnology for a Sustainable Future*; Springer Singapore, 2017.
45. Shah, M. P.; Ed.. *Removal of Emerging Contaminants Through Microbial Processes.* Singapore: Springer, 2021.
46. Sharma, P.; Parakh, S. K.; Singh, S. P.; Parra-Saldívar, R.; Kim, S. H.; Varjani, S.; Tong, Y. W. A Critical Review on Microbes-based Treatment Strategies for Mitigation of Toxic Pollutants. *Sci. Total Environ.* **2022,** 155444.
47. Spellman, F. R. *Spellman's Standard Handbook for Wastewater Operators: Intermediate Level*, Vol. II; Crc Press, 1999.
48. Preethi, P. S.; Hariharan, N. M.; Vickram, S.; Manian, R.; Manikandan, S.; Subbaiya, R.; Awasthi, M. K. Advances in Bioremediation of Emerging Contaminants from Industrial Wastewater by Oxidoreductase Enzymes. *Bioresour. Technol.* **2022,** 127444.
49. Kesari, K. K.; Soni, R.; Jamal, Q. M. S.; Tripathi, P.; Lal, J. A.; Jha, N. K.; Ruokolainen, J. Wastewater Treatment and Reuse: A Review of its Applications and Health Implications. *Water Air Soil Pollut.* **2021,** *232* (5), 1–28.

CHAPTER 4

IoT-Based Smart Water Leak Detection System for a Sustainable Future: A Case Study

KIAN HARIRI ASLI

Department of Electronic Engineering, University of Rome "Tor Vergata", Rome, Italy

ABSTRACT

Water leakage caused by broken pipes and connections in the water distribution network and water transmission line is identified and fixed after receiving a leak report from local residents. Generally, water leakage caused by network failures and incidents is considered visible types of leak. This type of leakage can be easily removed on the water systems. Water leakage due to underground faucets and fittings is considered to be part of invisible leaks. The minimum amount of water not taken into account due to faucets and fittings assumes important and must be calculated. Hence, this work aimed to implement advanced techniques for controlling invisible water leakage. The water distribution network and water transmission line operate reliably as a result of this procedure. The invisible leakage control is carried out by the improvement of remote sensing, networked sensors, modems, data loggers, Internet of things and GEO-database intercommunication process. Results showed that the invisible leaks were the major parts of unaccounted for water in the distribution network and water transmission line.

Sustainable Water Engineering: Smart and Emerging Technologies. Ali Pourhashemi, Kian Hariri Asli, Ann Rose Abraham, & A. K. Haghi (Eds.)

4.1 INTRODUCTION

The purpose of maintenance operation is the proper use of devices and equipment within their useful life. In other words, the manufactured devices should be used according to their special method and the manufacturing factory's instructions, and in a suitable period of time. They should have a suitable function and useful production with a minimum cost. If a device that is made for a certain function is not used for its intended purpose and more work is taken from it, it is certain that the device is not being used optimally, and it will not only cause damage but also will reduce its useful life.

Generally, a method is provided for every device that is made by its manufacturer to use it optimally as a start-up and operation manual. Nowadays, there is almost no device that comes without a manual. If the operation of a device is not followed according to its manufacturer's instructions, and the necessary repairs are not done on it, the mentioned device will be damaged. It will cause damage, and additional money should be spent for its restoration. In this case, the device is not used optimally and its useful life has been damaged.

The above content applies to all devices and equipment in a unit, and if a series of facilities are forced to work hard, their useful life will be greatly reduced. Because of this reason, it is necessary to operate and repair a facility. It should also be done according to the specific instructions of maintenance.

In the past, equipment repair was mainly done during production interruption due to equipment failure. The repair time is not certain for the maintenance personnel. The repairs should be done in such a way that their work is completed as soon as the production line is ready for work.

Obviously, this procedure cannot guarantee the satisfactory operation of the equipment, and therefore, in the long run, it not only causes a severe reduction in the lifespan of these equipment but also their premature failure in turn can cause an unwanted stop in the system and financial losses.[1–8]

In the last one or two decades, the concept of repair and maintenance has changed significantly. Today, repairs are divided into three categories:

- Corrective maintenance and repair.
- Repair and preventive maintenance.
- Repair and maintenance based on forecast.
- Corrective maintenance and repair.

Corrective maintenance consists of the same tasks that are usually known as maintenance. This type of repair is done only when the device in question is out of order or at least has problems and the relevant personnel have noticed this.

Preventive maintenance is completely different from the corrective maintenance. In this regard, the devices are periodically serviced in such a way as to prevent unexpected events and production stoppages due to device breakdowns. To achieve this goal, the following factors should be considered:

- Correct choice of electrical equipment.
- Choosing the right time interval for services and visits.
- Choosing the right time for basic repairs.
- The availability of materials and spare parts in the warehouse.
- Establishing the annual maintenance budget.
- Recording all the repair and maintenance work done in the device file.
- Repair and maintenance based on forecast.

This type of maintenance is actually a special form of preventive maintenance. This type of repair is done using precise tools, such as vibration meter, amplitude meter, pressure meter, thermometer, and resistance meters. These tools predict the problems before they occur and act to solve them.

The visits and services mentioned in this report are mainly related to preventive maintenance and to some extent predictive maintenance. Corrective maintenance and repair should be done when the desired equipment is out of order and the relevant personnel have not noticed the defect.[9–12]

4.1.1 OBJECTIVES

The following are among the reasons and goals that make it necessary to prepare and use the system management and maintenance program in urban water facilities.[13–22]

- Creating integrated management.
- Organizing, forming, and coordinating activities in the form of a coordinated system according to the needs of the distribution system.
- Increasing the efficiency and capabilities of the human force by improving their scientific and practical awareness along with the targeted control and guidance of the human force.
- Optimum use of existing capacities and facilities.

- Improving and upgrading the exploitation methods by increasing the exploitation power and economic maintenance of water supply networks and facilities, maximum use of the investments made by preventing their premature consumption.
- Preparation of regular and systematic methods of recording the information of the performed activities and improving the methods of obtaining and recording information in order to develop future policies related to maintenance and repairs.
- Creating a reliable system in providing services.
- Preventing the occurrence of problems caused by incomplete methods of inspection, maintenance, and repairs due to the traditional nature of the existing methods.
- Lack of inspection, repair, and maintenance programs of urban water facilities.
- Preventing sudden breakdowns in facilities, stopping services, reducing water loss, and time loss.
- Using special recipes for each device to use it.
- Modifying and replacing more appropriate methods, recipes, and tools to prevent similar defects from occurring.
- Drawing the curves and preparing the tables and comparing them with the operating curves of each device in order to identify the optimal condition, increase the quality of operation and profitability of the set of facilities by following the instructions.

4.1.2 PRINCIPLES OF EXPLOITATION

4.1.2.1 EDUCATION

All the operation, maintenance, and repair staff of a system should have sufficient mastery of the theoretical and practical aspects of this task, at least to the extent of the assigned task. Although, its usefulness is determined in the long run. In education, more attention should be paid to its practical aspects. The goals that can be achieved as a result of training are given as the following:

- Mastery of assigned duties and as a result optimal use of devices.
- Reduction of construction costs as a result of optimal operation.
- Increasing the probability of reaching the useful life of the system.

- Creating confidence in the use of the system, especially in the vital institutions of the society.
- Reducing risks as a result of awareness and correct use of devices in the system.

Maintenance of electrical systems and precision instruments should be done by people who have received sufficient training in the desired fields. Although practical training while working is very necessary, it cannot replace specialized professional technical training courses. Therefore, the personnel of electrical units and precision instruments must be familiar with the following during certain training courses:

- Safety regulations and rules for working with the electrical equipment.
- Basics of electricity and familiarity with production, distribution, and consumption systems of electric energy.
- Knowing different consumers and the characteristics of each of them.
- Being familiar with various equipment, including transformers, motors, keys, fuses, etc.
- Understanding the different circuits for starting and controlling of motors and other electrical equipment.
- Fire extinguishers in general.
- All tools and devices used in installation, start-up and repair, and maintenance of electrical equipment.

In addition to the above, the electrical personnel must be fully familiar with the with the following items:

- water supply process
- the electrical network
- the details of how the panels are connected to each other
- the transformer substation
- how consumers connect with the panels and control circuits
- their start-up during a training course inside the facilities

Obviously, the complete map of all electrical systems should be available to all of them.

During a training course, the personnel should also gain complete familiarity with the precision instrument of water supply process, the precision instruments installed in the units, the purpose of their installation and operation, and all the existing control circuits and precision instruments. It is obvious that the complete map of control circuits and precision instruments of all parts should be provided to them and they should be familiar with the following aspects:

- Basics of fluids and heat.
- Measuring equipment and precision instruments, their working methods and internal structure.
- Types of control systems, purpose, and application of each.
- Methods of calibrating precision instruments and devices.

Control in operation

In the operation of facilities, there are always important parameters that must be carefully controlled, such as the control of safety and security systems, the control of transmission and pumping line systems, and the control of electrical systems and precision instruments.

4.1.2.2 DAILY REPORT

A daily report of the operating system should contain the complete status of the systems and include parameters, such as height, speed, volume, working hours, work status, incidents and events, statistics, etc., during 24 working hours.

Based on these foundations and principles, it is possible to reach the goals that were mentioned at the beginning and make optimal use of the system according to the existing instructions:

- The criterion for the necessary work for the periodic services of each device is primarily the repair and maintenance manual of that device, which is printed by the manufacturer. Otherwise, it should be based on the current manual and the experience of repair personnel.
- In the works required for periodic services mentioned in this program, the mention of services and visits that should be done in shorter time intervals has been avoided. While their performance is necessary and the repair personnel should pay attention to this issue. For example, during monthly services, daily and weekly visits and services are also necessary.
- The periodic visits and services mentioned in the points taken from the equipment maintenance and repair manuals are mainly related to the normal working conditions of the equipment in general.

Therefore, different programs and components should be planned for the equipment that work in special conditions. For example, some of the different working conditions are explained below.

- Newly replaced devices need fewer visits and services than other devices.
- Devices that have worked for a long time and are worn out. It needs more visits and services and should be serviced in shorter intervals.
- Devices and equipment that are located in humid places or are exposed to dust and smoke, or installed inside the hall but near a door or window that is usually open, or the environment around them is corroded or exposed at higher temperatures, they should be serviced in shorter time intervals.

4.1.3 CASE STUDY; PILOT CITY OF THIS WORK

4.1.3.1 WATER DISTRIBUTION NETWORK MAPS

All the maps of the city's water distribution network at this work as a case study; which had not been reflected in the network, were updated with local survey and control of the places and accurate determination of the location of faucets, pipes, and pressure gauges (Figure 4.1).

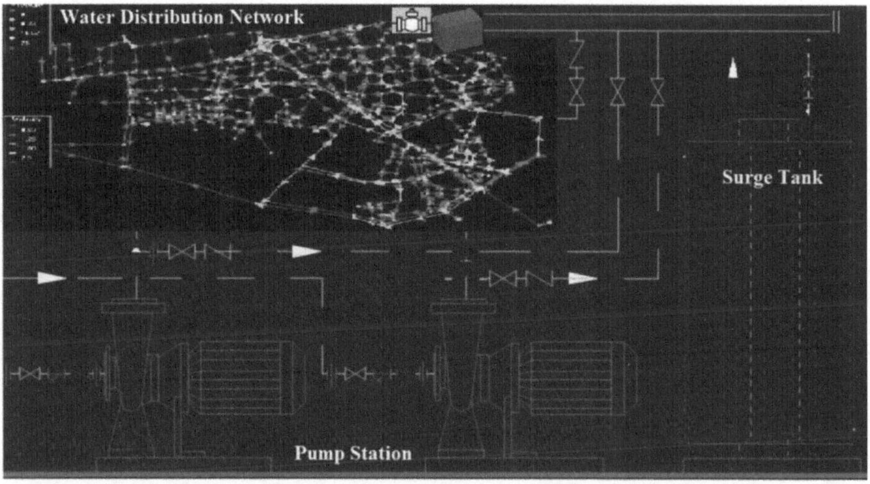

FIGURE 4.1 Maps of the city's water distribution network at this work.

These maps were scanned and edited, and all the problems of mapping and layering were solved. All computer maps were also updated. These maps were attached to the report as a separate album.

It was necessary to reflect the subsequent changes in the network and places on these maps so that they did not lose their effectiveness over time.

4.1.3.2 TECHNICAL SPECIFICATIONS OF FACILITIES AND EQUIPMENT

The facilities and equipment of Pilot City include the following (Figure 4.2):

- *Pipes:* In the computer maps, the pipes in the network were carefully updated and reflected on the maps by mentioning the type and diameter.
- *Faucets:* The location of these faucets, including visible faucets, invisible faucets, and all new faucets, were reflected on the water distribution network maps according to existing standards.
- The technical specifications of the facilities (Tables 4.1–4.20) of the pilot range were also given in a separate booklet under the title of "Network Faucets".[20–31]
- *Pressure gauges:* the location of these pressure gauges were reflected on the distribution network maps.
- Wells
- Chlorination station

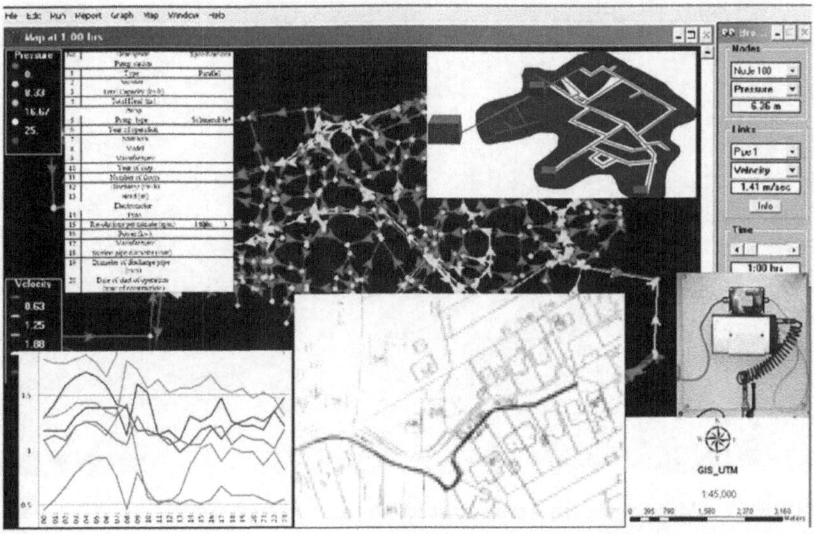

FIGURE 4.2 The facilities and equipment of pilot city.

TABLE 4.1 Technical Specifications of the Wells and Pumps (Deep Well; Submersible Pump; Low Speed).

No.	Equipment		Specifications
		Well number	
		Location	
		Well pipe diameter (mm)	
		Depth (m)	
1	Well	Static level	
		Discharge (m^3/h)	
		Amount of extractable water (L/s)	
		Type of well	Drilled well (deep)
		Dynamic level	
		Year of operation	
		Pump number	
		Purpose	Water displacement
		Number	
		Model	
		Manufacturer	
		Type	Submersible*
		Number of stages	
2	Pump	Capacity (m^3/h)	
		Suction pipe diameter (mm)	
		Discharge pipe diameter (mm)	
		Operating discharge (m^3/h)	
		Diameter of bilge pipe (mm)	
		Dynamic level (m)	
		Installation depth (m)	
		Injection site	
		Stop date	
		Type	
		Revolutions per minute (rpm)	(rpm) 1400
3	Electromotor	Power (kW)	
		Ampere (A)	
		Voltage (V)	
4		Explanations	

*A submersible pump is submerged in the water well or water reservoir. In this type of pump, the cavitation phenomena cannot be formed.

TABLE 4.2 Technical Specifications of the Wells and Pumps (Deep Well; Submersible Pump; High Speed).

No.	Equipment		Specifications
		Well number	
		Location	
		Well pipe diameter (mm)	
		Depth (m)	
1	Well	Static level	
		Discharge (m^3/h)	
		Amount of extractable water (L/s)	
		Type of well	Drilled well (deep)
		Dynamic level	
		Year of operation	
		Pump number	
		Purpose	Water displacement
		Number	
		Model	
		Manufacturer	
		Type	Submersible*
		Number of stages	
2	Pump	Capacity (m^3/h)	
		Suction pipe diameter (mm)	
		Discharge pipe diameter (mm)	
		Operating discharge (m^3/h)	
		Diameter of bilge pipe (mm)	
		Dynamic level (m)	
		Installation depth (m)	
		Injection site	
		Stop date	
		Type	
		Revolutions per minute (rpm)	(rpm) 2900
3	Electromotor	Power (kW)	
		Ampere (A)	
		Voltage (V)	
4		Explanations	

*A submersible pump is submerged in the water well or water reservoir. In this type of pump, the cavitation phenomena cannot be formed.

TABLE 4.3 Technical Specifications of the Wells and Pumps (Deep Well; Centrifugal Pump; Low Speed).

No.	Equipment		Specifications
		Well number	
		Location	
		Well pipe diameter (mm)	
		Depth (m)	
1	Well	Static level	
		Discharge (m^3/h)	
		Amount of extractable water (L/s)	
		Type of well	Drilled well (deep)
		Dynamic level	
		Year of operation	
		Pump number	
		Purpose	Water displacement
		Number	
		Model	
		Manufacturer	
		Type	Centrifugal*
		Serial number	
2	Pump	Capacity (m^3/h)	
		Suction pipe diameter (mm)	
		Discharge pipe diameter (mm)	
		Operating discharge (m^3/h)	
		Diameter of bilge pipe (mm)	
		Dynamic level (m)	
		Installation depth (m)	
		Injection site	
		Stop date	
		Type	
		Revolutions per minute (rpm)	(rpm) 1400
3	Electromotor	Power (kW)	
		Ampere (A)	
		Voltage (V)	
4		Explanations	

*A Centrifugal pump is used to transport water by conversion of the rotational kinetic energy to the hydrodynamic energy by the impellers and electromotor.

TABLE 4.4 Technical Specifications of the Wells and Pumps (Deep Well; Centrifugal Pump; High Speed).

No.	Equipment		Specifications
		Well number	
		Location	
		Well pipe diameter (mm)	
		Depth (m)	
1	Well	Static level	
		Discharge (m^3/h)	
		Amount of extractable water (L/s)	
		Type of well	Drilled well (deep)
		Dynamic level	
		Year of operation	
		Pump number	
		Purpose	Water displacement
		Number	
		Model	
		Manufacturer	
		Type	Centrifugal*
		Serial number	
2	Pump	Capacity (m^3/h)	
		Suction pipe diameter (mm)	
		Discharge pipe diameter (mm)	
		Operating discharge (m^3/h)	
		Diameter of bilge pipe (mm)	
		Dynamic level (m)	
		Installation depth (m)	
		Injection site	
		Stop date	
		Type	
		Revolutions per minute (rpm)	(rpm) 2900
3	Electromotor	Power (kW)	
		Ampere (A)	
		Voltage (V)	
4		Explanations	

*A Centrifugal pump is used to transport water by conversion of the rotational kinetic energy to the hydrodynamic energy by the impellers and electromotor.

TABLE 4.5 Technical Specifications of the Wells and Pumps (Shallow Well; Submersible Pump; Low Speed).

No.	Equipment		Specifications
		Well number	
		Location	
		Well pipe diameter (mm)	
		Depth (m)	
1	Well	Static level	
		Discharge (m^3/h)	
		Amount of extractable water (L/s)	
		Type of well	Driven well (shallow)
		Dynamic level	
		Year of operation	
		Pump number	
		Purpose	Water displacement
		Number	
		Model	
		Manufacturer	
		Type	Submersible*
		Number of stages	
2	Pump	Capacity (m^3/h)	
		Suction pipe diameter (mm)	
		Discharge pipe diameter (mm)	
		Operating discharge (m^3/h)	
		Diameter of bilge pipe (mm)	
		Dynamic level (m)	
		Installation depth (m)	
		Injection site	
		Stop date	
		Type	
		Revolutions per minute (rpm)	(rpm) 1400
3	Electromotor	Power (kW)	
		Ampere (A)	
		Voltage (V)	
4		Explanations	

*A submersible pump is submerged in the water well or water reservoir. In this type of pump, the cavitation phenomena cannot be formed.

TABLE 4.6 Technical Specifications of the Wells and Pumps (Shallow Deep Well; Submersible Pump; High Speed).

No.	Equipment		Specifications
		Well number	
		Location	
		Well pipe diameter (mm)	
		Depth (m)	
1	Well	Static level	
		Discharge (m^3/h)	
		Amount of extractable water (L/s)	
		Type of well	Driven well (Shallow)
		Dynamic level	
		Year of operation	
		Pump number	
		Purpose	Water displacement
		Number	
		Model	
		Manufacturer	
		Type	Submersible*
		Number of stages	
2	Pump	Capacity (m^3/h)	
		Suction pipe diameter (mm)	
		Discharge pipe diameter (mm)	
		Operating discharge (m^3/h)	
		Diameter of bilge pipe (mm)	
		Dynamic level (m)	
		Installation depth (m)	
		Injection site	
		Stop date	
		Type	
		Revolutions per minute (rpm)	(rpm) 2900
3	Electromotor	Power (kW)	
		Ampere (A)	
		Voltage (V)	
4		Explanations	

*A submersible pump is submerged in the water well or water reservoir. In this type of pump, the cavitation phenomena cannot be formed.

TABLE 4.7 Technical Specifications of the Wells and Pumps (Shallow Well; Centrifugal Pump; Low Speed).

No.	Equipment		Specifications
1	Well	Well number	
		Location	
		Well pipe diameter (mm)	
		Depth (m)	
		Static level	
		Discharge (m^3/h)	
		Amount of extractable water (L/s)	
		Type of well	Driven well (Shallow)
		Dynamic level	
		Year of operation	
2	Pump	Pump number	
		Purpose	Water displacement
		Number	
		Model	
		Manufacturer	
		Type	Centrifugal*
		Serial number	
		Capacity (m^3/h)	
		Suction pipe diameter (mm)	
		Discharge pipe diameter (mm)	
		Operating discharge (m^3/h)	
		Diameter of bilge pipe (mm)	
		Dynamic level (m)	
		Installation depth (m)	
		Injection site	
		Stop date	
3	Electromotor	Type	
		Revolutions per minute (rpm)	(rpm) 1400
		Power (kW)	
		Ampere (A)	
		Voltage (V)	
4		Explanations	

*A centrifugal pump is used to transport water by conversion of the rotational kinetic energy to the hydrodynamic energy by the impellers and electromotor.

TABLE 4.8 Technical Specifications of the Wells and Pumps (Shallow Well; Centrifugal Pump; High Speed).

No.	Equipment		Specifications
		Well number	
		Location	
		Well pipe diameter (mm)	
		Depth (m)	
1	Well	Static level	
		Discharge (m^3/h)	
		Amount of extractable water (L/s)	
		Type of well	Driven well (Shallow)
		Dynamic level	
		Year of operation	
		Pump number	
		Purpose	Water displacement
		Number	
		Model	
		Manufacturer	
		Type	Centrifugal*
		Serial number	
2	Pump	Capacity (m^3/h)	
		Suction pipe diameter (mm)	
		Discharge pipe diameter (mm)	
		Operating discharge (m^3/h)	
		Diameter of bilge pipe (mm)	
		Dynamic level (m)	
		Installation depth (m)	
		Injection site	
		Stop date	
		Type	
		Revolutions per minute (rpm)	(rpm) 2900
3	Electromotor	Power (kW)	
		Ampere (A)	
		Voltage (V)	
4		Explanations	

*A centrifugal pump is used to transport water by conversion of the rotational kinetic energy to the hydrodynamic energy by the impellers and electromotor.

TABLE 4.9 Technical Specifications of the Wells and Pumps (Shallow Well; Submersible Pump; Low Speed).

No.	Equipment		Specifications
		Well number	
		Location	
		Well pipe diameter (mm)	
		Depth (m)	
1	Well	Static level	
		Discharge (m^3/h)	
		Amount of extractable water (L/s)	
		Type of well	Dug well (shallow)
		Dynamic level	
		Year of operation	
		Pump number	
		Purpose	Water displacement
		Number	
		Model	
		Manufacturer	
		Type	Submersible*
		Number of stages	
2	Pump	Capacity (m^3/h)	
		Suction pipe diameter (mm)	
		Discharge pipe diameter (mm)	
		Operating discharge (m^3/h)	
		Diameter of bilge pipe (mm)	
		Dynamic level (m)	
		Installation depth (m)	
		Injection site	
		Stop date	
		Type	
		Revolutions per minute (rpm)	(rpm) 1400
3	Electromotor	Power (kW)	
		Ampere (A)	
		Voltage (V)	
4		Explanations	

*A submersible pump is submerged in the water well or water reservoir. In this type of pump, the cavitation phenomena cannot be formed.

TABLE 4.10 Technical Specifications of the Wells and Pumps (Shallow Well; Submersible Pump; High Speed).

No.	Equipment		Specifications
		Well number	
		Location	
		Well pipe diameter (mm)	
		Depth (m)	
1	Well	Static level	
		Discharge (m^3/h)	
		Amount of extractable water (L/s)	
		Type of well	Dug well (shallow)
		Dynamic level	
		Year of operation	
		Pump number	
		Purpose	Water displacement
		Number	
		Model	
		Manufacturer	
		Type	*Submersible
		Number of stages	
2	Pump	Capacity (m^3/h)	
		Suction pipe diameter (mm)	
		Discharge pipe diameter (mm)	
		Operating discharge (m^3/h)	
		Diameter of bilge pipe (mm)	
		Dynamic level (m)	
		Installation depth (m)	
		Injection site	
		Stop date	
		Type	
		Revolutions per minute (rpm)	(rpm) 2900
3	Electromotor	Power (kW)	
		Ampere (A)	
		Voltage (V)	
4		Explanations	

*A submersible pump is submerged in the water well or water reservoir. In this type of pump, the cavitation phenomena cannot be formed.

TABLE 4.11 Technical Specifications of the Wells and Pumps (Shallow Well; Centrifugal Pump; Low Speed).

No.	Equipment		Specifications
		Well number	
		Location	
		Well pipe diameter (mm)	
		Depth (m)	
1	Well	Static level	
		Discharge (m^3/h)	
		Amount of extractable water (L/s)	
		Type of well	Dug well (shallow)
		Dynamic level	
		Year of operation	
		Pump number	
		Purpose	Water displacement
		Number	
		Model	
		Manufacturer	
		Type	Centrifugal*
		Serial number	
2	Pump	Capacity (m^3/h)	
		Suction pipe diameter (mm)	
		Discharge pipe diameter (mm)	
		Operating discharge (m^3/h)	
		Diameter of bilge pipe (mm)	
		Dynamic level (m)	
		Installation depth (m)	
		Injection site	
		Stop date	
		Type	
		Revolutions per minute (rpm)	(rpm) 1400
3	Electromotor	Power (kW)	
		Ampere (A)	
		Voltage (V)	
4		Explanations	

*A centrifugal pump is used to transport water by conversion of the rotational kinetic energy to the hydrodynamic energy by the impellers and electromotor.

TABLE 4.12 Technical Specifications of the Wells and Pumps (Shallow Well; Centrifugal Pump; High Speed).

No.	Equipment		Specifications
		Well number	
		Location	
		Well pipe diameter (mm)	
		Depth (m)	
1	Well	Static level	
		Discharge (m^3/h)	
		Amount of extractable water (L/s)	
		Type of well	Dug well (shallow)
		Dynamic level	
		Year of operation	
		Pump number	
		Purpose	Water displacement
		Number	
		Model	
		Manufacturer	
		Type	Centrifugal*
		Serial number	
2	Pump	Capacity (m^3/h)	
		Suction pipe diameter (mm)	
		Discharge pipe diameter (mm)	
		Operating discharge (m^3/h)	
		Diameter of bilge pipe (mm)	
		Dynamic level (m)	
		Installation depth (m)	
		Injection site	
		Stop date	
		Type	
		Revolutions per minute (rpm)	(rpm) 2900
3	Electromotor	Power (kW)	
		Ampere (A)	
		Voltage (V)	
4		Explanations	

*A centrifugal pump is used to transport water by conversion of the rotational kinetic energy to the hydrodynamic energy by the impellers and electromotor.

TABLE 4.13 Technical Specifications of Pump Stations (Series Pumps; Submersible Pumps; Low Speed).

No.	Description	Specifications
	Pump station	
1	Type	Series
2	Number	
3	Capacity (kg/h) total	
4	Total Head (m)	
	Pump	
5	Type of pump	Submersible*
6	Year of operation	
7	Numbers	
8	Model	
9	Manufacturer	
10	Year of stop	
11	Number of floors	
12	Discharge (m^3/h)	
13	Head (m)	
	Electromotor	
14	Type	
15	Revolutions per minute (rpm)	(rpm) 1400
16	Power (kW)	
17	Manufacturer	
18	Suction pipe diameter (mm)	
19	Diameter of discharge pipe (mm)	
20	Date of start of operation (year of construction)	

*A submersible pump is submerged in the water well or water reservoir. In this type of pump, the cavitation phenomena cannot be formed.

TABLE 4.14 Technical Specifications of Pump Stations (Series Pumps; Submersible Pumps; High Speed).

No.	Description	Specifications
	Pump station	
1	Type	Series
2	Number	
3	Capacity (kg/h) total	
4	Total Head (m)	
	Pump	
5	Type of pump	Submersible*
6	Year of operation	
7	Numbers	
8	Model	
9	Manufacturer	
10	Year of stop	
11	Number of floors	
12	Discharge (m^3/h)	
13	Head (m)	
	Electromotor	
14	Type	
15	Revolutions per minute (rpm)	(rpm) 2900
16	Power (kW)	
17	Manufacturer	
18	Suction pipe diameter (mm)	
19	Diameter of discharge pipe (mm)	
20	Date of start of operation (year of construction)	

*A submersible pump is submerged in the water well or water reservoir. In this type of pump, the cavitation phenomena cannot be formed.

TABLE 4.15 Technical Specifications of Pump Stations (Parallel Pumps; Submersible Pumps; Low Speed).

No.	Description	Specifications
	Pump station	
1	Type	Parallel
2	Number	
3	Capacity (kg/h) total	
4	Total Head (m)	
	Pump	
5	Type of pump	Submersible*
6	Year of operation	
7	Numbers	
8	Model	
9	Manufacturer	
10	Year of stop	
11	Number of floors	
12	Discharge (m^3/h)	
13	Head (m)	
	Electromotor	
14	Type	
15	Revolutions per minute (rpm)	(rpm) 1400
16	Power (kW)	
17	Manufacturer	
18	Suction pipe diameter (mm)	
19	Diameter of discharge pipe (mm)	
20	Date of start of operation (year of construction)	

*A submersible pump is submerged in the water well or water reservoir. In this type of pump, the cavitation phenomena cannot be formed.

TABLE 4.16 Technical Specifications of Pump Stations (Parallel Pumps; Submersible Pumps; High Speed).

No.	Description	Specifications
	Pump station	
1	Type	Parallel
2	Number	
3	Capacity (kg/h) total	
4	Total head (m)	
	Pump	
5	Type of pump	Submersible*
6	Year of operation	
7	Numbers	
8	Model	
9	Manufacturer	
10	Year of stop	
11	Number of floors	
12	Discharge (m^3/h)	
13	Head (m)	
	Electromotor	
14	Type	
15	Revolutions per minute (rpm)	(rpm) 2900
16	Power (kW)	
17	Manufacturer	
18	Suction pipe diameter (mm)	
19	Diameter of discharge pipe (mm)	
20	Date of start of operation (year of construction)	

*A submersible pump is submerged in the water well or water reservoir. In this type of pump, the cavitation phenomena cannot be formed.

TABLE 4.17 Technical Specifications of Pump Stations (Series Pumps; Centrifugal Pumps; Low Speed).

No.	Description	Specifications
	Pump station	
1	Type	Series
2	Number	
3	Capacity (kg/h) total	
4	Total head (m)	
	Pump	
5	Type of pump	Centrifugal*
6	Year of operation	
7	Numbers	
8	Model	
9	Manufacturer	
10	Year of stop	
11	Number of floors	
12	Discharge (m^3/h)	
13	Head (m)	
	Electromotor	
14	Type	
15	Revolutions per minute (rpm)	(rpm) 1400
16	Power (kW)	
17	Manufacturer	
18	Suction pipe diameter (mm)	
19	Diameter of discharge pipe (mm)	
20	Date of start of operation (year of construction)	

*A centrifugal pump is used to transport water by conversion of the rotational kinetic energy to the hydrodynamic energy by the impellers and electromotor.

TABLE 4.18 Technical Specifications of Pump Stations (Series Pumps; Centrifugal Pumps; High Speed).

No.	Description	Specifications
	Pump station	
1	Type	Series
2	Number	
3	Capacity (kg/h) total	
4	Total Head (m)	
	Pump	
5	Type of pump	Centrifugal*
6	Year of operation	
7	Numbers	
8	Model	
9	Manufacturer	
10	Year of stop	
11	Number of floors	
12	Discharge (m^3/h)	
13	Head (m)	
	Electromotor	
14	Type	
15	Revolutions per minute (rpm)	(rpm) 2900
16	Power (kW)	
17	Manufacturer	
18	Suction pipe diameter (mm)	
19	Diameter of discharge pipe (mm)	
20	Date of start of operation (year of construction)	

*A centrifugal pump is used to transport water by conversion of the rotational kinetic energy to the hydrodynamic energy by the impellers and electromotor.

TABLE 4.19 Technical Specifications of Pump Stations (Parallel Pumps; Centrifugal Pumps; Low Speed).

No.	Description	Specifications
	Pump station	
1	Type	Parallel
2	Number	
3	Capacity (kg/h) total	
4	Total head (m)	
	Pump	
5	Type of pump	Centrifugal*
6	Year of operation	
7	Numbers	
8	Model	
9	Manufacturer	
10	Year of stop	
11	Number of floors	
12	Discharge (m^3/h)	
13	Head (m)	
	Electromotor	
14	Type	
15	Revolutions per minute (rpm)	(rpm) 1400
16	Power (kW)	
17	Manufacturer	
18	Suction pipe diameter (mm)	
19	Diameter of discharge pipe (mm)	
20	Date of start of operation (year of construction)	

*A centrifugal pump are used to transport water by conversion of the rotational kinetic energy to the hydrodynamic energy by the impellers and electromotor.

TABLE 4.20 Technical Specifications of Pump Stations ((Parallel Pumps; Centrifugal Pumps; High Speed).

No.	Description	Specifications
	Pump station	
1	Type	Parallel
2	Number	
3	Capacity (kg/h) total	
4	Total Head (m)	
	Pump	
5	Type of pump	Centrifugal*
6	Year of operation	
7	Numbers	
8	Model	
9	Manufacturer	
10	Year of stop	
11	Number of floors	
12	Discharge (m^3/h)	
13	Head (m)	
	Electromotor	
14	Type	
15	Revolutions per minute (rpm)	(rpm) 2900
16	Power (kW)	
17	Manufacturer	
18	Suction pipe diameter (mm)	
19	Diameter of discharge pipe (mm)	
20	Date of start of operation (year of construction)	

*A centrifugal pump are used to transport water by conversion of the rotational kinetic energy to the hydrodynamic energy by the impellers and electromotor.

Water loss at the pilot city of this work

$$\text{Water Loss} = (\text{Quantity of water put in to supply} - (\text{Non-domestic usage} + \text{Domestic consumption})) \tag{4.1}$$

Unaccounted for water

$$U = S - (M + (A * P)) \tag{4.2}$$

where

- U: Unknown or unaccounted for quantities of water including leakage.
- S: Sum of all water inputs into a system.
- M: Sum of all water accounted for by measure (metered supplies, domestic and non-domestic)
- A: Average domestic usage per capita of population.
- P: Population supplied (non-metered).

4.1.3.3 UNACCOUNTED FOR WATER (UFW) IN THE DISTRIBUTION NETWORK

Unaccounted for water (UFW) in the water distribution network includes the following situations:

- Water losses due to fractures and incidents (visible leakage).
- Water losses due to small leaks (invisible leaks).
- Water losses from faucets and fittings.
- Removal of fire extinguishers.
- Washing the network through the available drain valves.

Small leaks caused by faucets and fittings are considered to be part of small or invisible leaks. In this way, the minimum amount of water that is not taken into account due to faucets and fittings can be calculated.[21,32–38]

Since leaks caused by broken pipes and connections in the water distribution network are identified and fixed after receiving a leak report from the local residents, leaks caused by network failures and incidents are considered visible types of leaks.

Regarding the leakage caused by the removal of fire hydrants, according to the investigations, this consumption is supplied through the wells that have been allocated for this purpose, and it has not been included in the calculations of the leakage from the network (Table 4.21).

Water balance method of computing leakage

Total integrated flow formula:

$$U = S - (M + A * P) \tag{4.3}$$

where U: Unaccounted for water S: Total volume supplied M: Metered use A: Per capita use P: Population supplied unmetered

The background night flow losses (when no bursts exist in a DMA) can be calculated for any DMA; given L (length of mains in km), N (number of properties), and AZNP (average zonal night pressure in m)) from eq (4.4):

$$NFLB\left(\frac{l}{hr}\right) = \left(C1 * L + \left(C2 + C3\right) * N\right) * PCF \tag{4.4}$$

(Based on the values of C1, C2, and C3 from Table 4.21, and the pressure correction factor (PCF) from Table 4.22.)

TABLE 4.21 Background Night Flow Losses.

Background losses component	Units	Low	Average	High
C1:	L/KM/HR	20	40	60
C2:	L/PROP/HR	1.5	3	4.5
C3:	L/PROP/HR	.5	1	1.5

TABLE 4.22 Pressure Correction Factors.

AZPN (m)	20	30	40	50	60	70	80	90	100
PCF	.329	.529	.753	1	1.271	1.565	1.884	2.226	2.592

4.1.4 OPERATION AND MAINTENANCE AT THE PILOT CITY OF THIS WORK

4.1.4.1 EDUCATION

The training of maintenance for repair personals is useful and important. The correct operation directly affects the work of the maintenance group, and the correct repair and maintenance of the system increases the efficiency of the operation group.

Control in repair and maintenance

The maintenance of each system is done in three ways:

- continuous control of system details
- periodic services
- periodic replacements

In general, the time interval of each service or replacement is determined by the instructions of the manufacturer, but the maintenance manager has interfered in these instructions by changing the operating conditions in such a way that the efficiency of the system is maintained at the optimal level.

Regarding the continuous control of the details of the system, the work of the maintenance and repair group is somewhat similar to the work of the operation group, with the difference that the maintenance and repair group, after observing the problems of the system, follow-up the investigations related to fault finding and fix them with more accuracy, usually comment on these cases with speculation.

Periodic services should be performed in each of the components of the system according to the manufacturer's instructions. In this way, if there are problems in any of the components of the system, before these cases cause serious problems in service or interruption of services, periodic replacements are also done based on the useful life of each device or part.

The useful life of each part is determined by taking into account the conditions of operation and exploitation of the part, so that in some cases, we see that some parts work more or less than the expected useful life.

However, although periodic replacements may be expensive in the first place, in order to prevent consequences that are sometimes much costlier, it is a reasonable and economical way to replace each part at the end of its useful life.

The totality of the principles mentioned in operation, repairs, and maintenance includes the following steps for each of the components of urban facilities.

Pumping stations

- Normal operation

  In operating the pumping station, the following items must be controlled:

Storage tanks (Figure 4.3)

- Water level control inside the tanks
- Checking the condition of the inlet and outlet valves

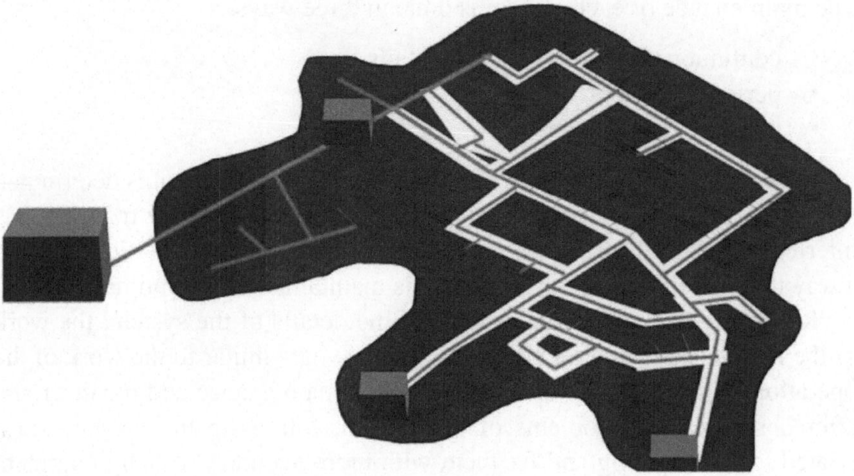

FIGURE 4.3 Storage tanks for the water distribution network.

Electric motors

- Engine sound control.
- Shell temperature control.
- Control of starting ampere, working ampere, voltage and engine speed.
- Electronic consumption registration.
- The output cables from the main switches and the starting switches of the electropumps are touched by hand, if the heat is more than reasonable 30°C, the switch should be turned off and the repair group should be informed.
- Cleaning the inside of the electric motor with a dry air compressor with a pressure of about 4 atm.

Pumps (Figure 4.4)

- Control of couplings.
- Controlling the sound of the pump and its vibration.

- Pump shell temperature control.
- Control the circulation of the cooling system, if any.
- Ventilation system control.
- Control of input and output current intensity.
- Suction pipeline
- Line negative pressure control.
- Checking the status of the valves on and off.
- Checking the sealing of valves and fittings.

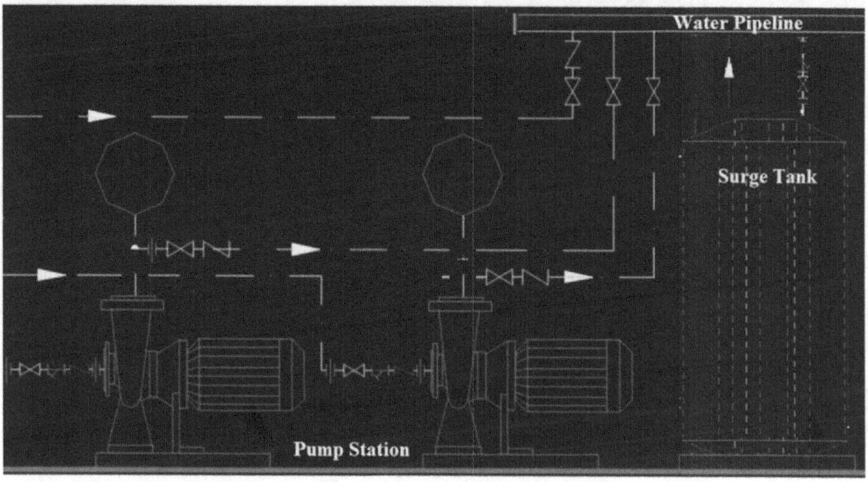

FIGURE 4.4 Pump station for the water distribution network.

Water pipeline (Figure 4.5)

- Positive line pressure control.
- Checking the status of faucets and connections.
- Sealing control of faucets and fittings.

Security and lighting system

- Controlling the status of security systems in the facilities.
- Controlling the status of the whole system by setting up a daily report.
- The lighting of the pump station during the day, control and if there is a defect, take action to fix it so that there is no problem at night.
- Fault finding and fixing it.

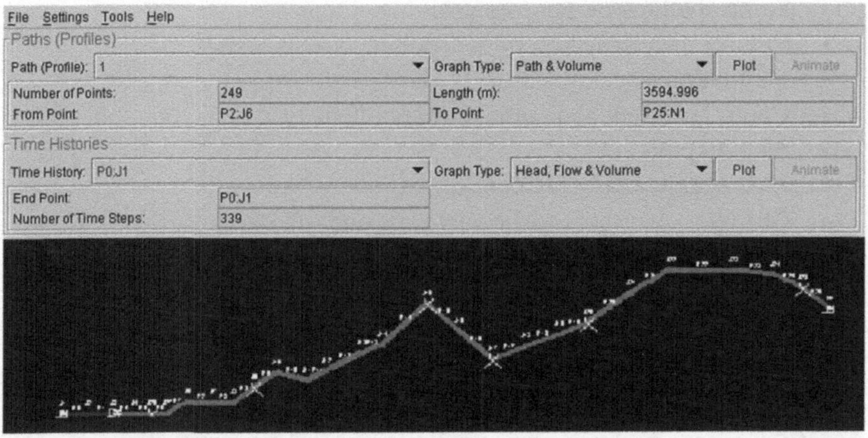

FIGURE 4.5 Water pipeline components.

If periodic services and repairs are carried out according to the schedule, the workload of the repair unit will be greatly reduced. In any case, there are reasons to form an experienced construction group at the pumping station to fix the defects and prevent disruptions in operation.

Defects and faults created in the electric motor of the pumps can have hydraulic or mechanical reasons. In the first case, the pump does not have the ability to pump, or the pump operates with less than its nominal capacity, or does not provide enough pressure; in the second case, the amount of electricity consumed increases and problems, such as the value of the pump cause noise, corrosion, and breakage of parts. In general, any defects and problems in the operation of pumps can be caused by one or more reasons. By finding the cause, the repair department manager can easily determine the percentage of its resolution.[4–6,39–42]

4.1.5 MAINTENANCE AND LUBRICATION

4.1.5.1 MAINTENANCE PRINCIPLES OF PUMPING STATIONS

The maintenance of a pumping station, like any industrial system, is divided into three parts:

- Continuous control of system details
- Periodic services
- Period replacements

In a pumping station, the details of the station must be continuously controlled by the maintenance group.

4.1.5.2 SYSTEM FAULTS

Periodical services of a pump station for each component usually include the following:

- Operations on valves and controlling one-way valves.
- Greasing the bearings of electric motors and pumps.
- Opening and closing the pumps and checking the condition of the shaft, impeller, and internal wall of the pump shell.
- For every ultimate working hours of the pump, the grease of the bearings should be replaced.
- Testing the start-up amp–work amp head of electric motors.
- Service of electrical panels and sensitivity testing of safety and protective circuits.
- Service of control systems, such as limit, switches, flowmeters.
- Service and control of loose couplings.

Periodic replacements of a pumping station for each of the components in a way that can include the following:

- Bearings and roller bearings (pumps and electric motors)
- Pump propellers
- Shaft of pumps
- Ball shaft of valves
- Manometers
- Couplings
- Water flow measurement contours
- Sensitive parts of electrical panels

4.1.6 PERIODIC OPERATION OF INSPECTION AND MAINTENANCE OF THE PUMPING STATION

4.1.6.1 DAILY OPERATION

All defects that are observed by the operators during hourly and daily visits in the operation of the pumps, such as a change in the sound of the pump motor, a sudden change in the temperature of the pump shell, leakage from the pump, pressure and flow readings, etc. to be registered.

4.1.6.2 MONTHLY OPERATION

In addition to the daily visits by the operators, the following visits and operations must be performed by the construction team on a monthly basis.

- Investigating the condition of the bolt and gland.
- Check the state of packing and sealing gaskets and replace them if necessary.
- Adjusting and aligning the pump motor, or in other words, inspecting and measuring the distance of the couplings.
- Greasing and lubrication of electric motor bearings with fireproof grease.
- Inspection and repair of pipes and connections if needed.
- Inspection and repair or replacement of pressure gauges if necessary.
- Pump electromotor vibration control.
- Cleaning the inside of the electric motor of the air compressor.

4.1.6.3 ANNUALLY OPERATION

The measures that should be taken during the annual visits are as follows:

- Pump components should be opened, washed, and cleaned.
- Descaling of impeller and pump shaft.
- Lubrication of bearings and roller bearings.
- Painting faucets, fittings, and pump shell.
- Inspect the floor and walls of the pump station and repair it if any defects are found.
- Close the valve located on the water pipeline of the desired pump.
- Turn off the engine.
- Close the valve located on the suction pipeline.
- Safety.
- The electropump should never be used outside the specified range in terms of water flow, motor speed, pressure, and temperature.
- Avoid keeping the electropump on for a long time while the outlet valve is closed.
- Under no circumstances should the pump work no load.

4.1.6.4 FAUCETS OF THE DISTRIBUTION NETWORK

The valves in the urban water distribution network are considered as key and control points of the network. The absence of a suitable plan and method for the maintenance and operation of the faucets of the distribution network often causes them to lose their desired and necessary performance after a period of installation. Instead of being used as executive management tools, it causes problems. They are distributed in the system. The first step to repair and maintain these units is to know their function and role in the distribution network. After that the existing obstacles should be removed and their protection and maintenance should be known, and then with the schedule of periodic visits and service, while applying the necessary controls, repair, and maintenance should be considered.

4.1.6.5 GATE VALVE

This type of valve is used to completely shut off and connect the water flow in the distribution network. These valves are manufactured and marketed in different sizes and to withstand different pressures. The material of these faucets is generally cast iron or ductile iron, and is widely used in distribution networks and water transmission lines. The most common installation locations of valves are:

- In the place of branches and the intersection of pipelines in order to disconnect and connect the water flow of the desired branch or the connection between the lines.
- At the entrance and exit of tanks.
- In water transmission lines known as taps.
- There are two types of valve in terms of how it works.
- A valve with a rising stem (rising stem valve) in which the stem or rod of the valve also rises and water is allowed to pass.
- A valve with a fixed stem (non-rising stem valve) is in this group, where the valve moves on a rod or stem. These valves are generally used for underground installation.[7–18, 33]

Advantages of valves

- When it is completely open, the water pressure drop is very small and can be ignored.

- Long lifespan.
- High work pressure.
- There is less water leakage in them.

Disadvantages of valves

- In large diameters and for high pressures, it is difficult to open and close it by hand, and the valve must have a passage.
- If the valve is not operated (open and close) regularly, due to the accumulation and deposition of particles in the water, opening and closing them will not be done easily and correctly.

4.1.6.6 REPAIRS

Valve valves often need to be repaired due to the following reasons.

- Deterioration of the sealing thread (packing).
- Corrosion and rusting of the tap rod or stem.
- Corrosion of casing or disc.
- Deterioration of sealing rings.

Each of the above causes malfunctions and leakage from the valves. In order to fix the defect and repair each of the above-mentioned items, it is done in the following way:

Replacing the sealing thread

- Open the valve gland.
- The sealing thread must be removed from the sealing chamber (suffing box) and its place must be cleaned.
- The thread of the new seal should be placed in its place and secured in place with a gland.
- Pour a few drops of oil on the sealing thread and then put the valve gland in its place and close it tightly.

The service of the bar or stem of valve

- Open the milk valve.
- Take out the sealing thread.
- Open the cap and remove its rod.

- Clean the surface of the rod, and if there is a lot of corrosion and rust, the rod should be replaced.
- Put the rod inside the cap.
- Seal gaskets should be replaced.
- The cap should be firmly in place.
- Replace the water seal thread.
- Open and close the valve.

Fixing the corrosion of the disk or valve casing and sealing ring

- Open the gland and cap.
- Remove the rod.
- Take out the disk and clean it and replace the disk if there is a lot of corrosion.
- Clean the sealing rings and replace them otherwise.
- Mount the valve parts again.
- Sealing gaskets and sealing thread should be replaced.

4.1.6.7 KEEPING

Valves should not be left open or closed for a long time. Because the water inside the valve is filled with salts when it is open, and the valve is not completely closed and sealed. In the case of being closed for a long time, the solutes in the water accumulate around the valve drawer and cause a lot of pressure on the valve axis when opening the valve, in which case there will be a possibility of the axis being cut. According to the above phrases, it is necessary to operate the valve at least once a month.

4.1.6.8 ONE-WAY CHECK VALVE

A one-way valve is actually a valve that is installed in the pipeline and when water flows, the moving force of the water opens it. When the water flow changes the direction, the plate closes on a tight seat. In general, one-way valves work automatically and prevent reverse flow in pipelines.

Hinge one-way valves are often used when the goal is to pass the flow completely, and it is the most common and most used one-way valve. The one-way valve is used in cases where the valve must operate frequently, and in networks with high pressure and high flow rate.

Among the factors that cause defects and disturbances in the performance of the one-way valve are corrosion, wear of the retaining pin, retaining cap, sedimentation of salts in the valve and corrosion of the disc. In order to fix such defects, proceed as follows:

- Open the valve cover.
- The nut or spine of the axis should be removed.
- The valve disc should be sanded or ground.
- In case of damage to the axle, the hub and nut of the axle and their cover should be replaced.
- Mount the valve components again.
- Valve should be tested.

4.1.6.9 BUTTERFLY VALVE

The construction of butterfly valves is simpler than other valves. These valves have a part in the shape of a rotating plate or disk that moves around an axis or a rod or a stem. Opening and closing these valves is simple and is done only by turning the disk so that it fills the entire width of the pipe. Depending on its type, the plastic lining of the seat of these valves is made of adhesive material, or it is fixed to the disc with mechanical devices, or it is embedded in the disc, or it is made of a mechanical spring seal, or a metal seat that is placed in the form of an O-RING (0-RING) around the edge of the disc.

The advantages of these valves are:

- The coefficient of water passage in it is directly related to the opening percentage of the valve, and therefore, it can be used to control the flow and create a pressure drop in the network.
- Opening and closing it is done with more ease and speed.
- It needs less space for installation.
- Its repair and maintenance cost is relatively low.

These valves are easy to maintain and generally do not need lubrication until the stem O-ring is replaced. The disc and stem as well as the O-ring disc and its plastic seats are easily replaceable. Replacement of the plastic metal lining of the sealing surfaces with adhesive materials cannot be done on site and is done only by the manufacturer.

4.1.6.10 AIR VALVE

An air valve is a device to let air out of a pipeline or to let air into it. Air valves are divided into three categories:

- Single-hole air valves, with a small hole, which is used to release air from the pipeline that is under hydraulic pressure.
- Single-orifice air valves, with a large orifice: This valve is used to ventilate and empty the pipe. When the pipeline is filled with water, air is removed from it, and when the pipe is emptied of water, air enters the pipeline. When the pipeline is under hydraulic pressure, the above valve is not able to release the small amount of collected air, and in this sense, the hydraulic pressure prevents the ball from separating from the hole.
- Two-hole air valve: This performs the work of both the above valves.
- In terms of operation, if the valves are installed in the suitable condition, they will not have any particular problem. Of course, it is necessary to visit it periodically, and if its opening is clogged with sediment, it should be descaled and rusted.

4.1.6.11 PERIODIC VISITS, REPAIR AND REPLACEMENT OF VALVES

Due to the fact that during the use of the distribution network, there is no justification for cutting off the water for a long time, and sometimes, it will cause protests from the consumers. Therefore, it is not enough to fix problems as corrective maintenance. Rather, it is necessary to optimize the use of the facilities by training and equipping the crews and conducting periodic inspections and services on time (while preventing the creation of a crisis).

Each of the valves used in the network, even if they are made of the best metal materials, will corrode, rust, and deposit with time.

Like other mechanical systems, failure to operate the valve on time will gradually make it impossible to maneuver the valve. So their operation is recommended twice a year.

On the other hand, the access valve to the line valve due to the spread and various implementation operations in the city, such as drilling, surface changes, and asphalt may be invisible, intentionally or unintentionally, and as a result, the possibility of valve operation is lost. The more important point is that each faucet is installed for a reason and to connect and disconnect the water in a specific area. If a pipe breaks due to drilling, leakage, or as a result of any kind of accident, by closing the valve of the relevant line, the outflow of water from the accident site is controlled. Therefore, if one of the valves in the network becomes invisible, at least several other line valves should be closed instead, which will cause the water to be cut off in a wider area, and water will be wasted for a long time.

In addition to the mentioned points, if any of the valves in the line is not opened and closed continuously, it gets corroded. It will automatically lose its water tightness after an operation and will leak from the outside, which results in water loss, water infiltration. As a result, the soil settles under the pipe and breaks the pipelines. If there is no sealing valve, the repair of the accident pipe is impossible, and in some cases, a pump is needed to drain the water from the accident site.[21,32,43–47]

4.1.6.12 WATER DISTRIBUTION NETWORK PIPES

The skeleton of the distribution network consists of pipes with small and large trains and relatively the same. These pipes, usually can be used in sizes of 60, 80, 100, and 150 mm for subsidiary networks and up to 1250 mm diameter for main networks. These pipes are the only part of the urban water network that are always buried underground, so they are more vulnerable than other parts of the network and will have accidents. Therefore, maintenance of the network in the sense of visiting them is not possible only when accidents occur, it is better to carry out leak detection by a leak detection device during a regular program and according to the priority of the areas according to the age of the network and its importance. The network should be repaired before causing any massive and costly damage. In this regard, it is necessary to mention that usually, the amount of water wasted from small leaks in the network is more than large leaks that cause accidents.

The reasons for reducing the water transfer capacity of pipes are the following:

- Foreign objects inside the pipes, such as wood, stone, which enter the pipes during pipe laying or necessary repairs.
- Sedimentation of sludge, iron, manganese and algae growth.
- Creating a coating layer of sedimentary materials on the inner wall of pipes and protrusions caused by rusting or water corrosion or lime deposits.
- Air accumulation in the highest points of the level.
- Reducing the cross-section of the pipe due to sedimentation of faucets in semi-closed states or bending of pipes.

Therefore, in order to prevent large losses, the accurate implementation of the network and the use of quality materials and the careful monitoring of its implementation while observing the technical principles of implementation are factors that can have a great effect on the health of the network and increase its life.

Otherwise, it is possible that the water transmission pipelines are reduced, the following measures should be taken:

- Full air evacuation control throughout the length of the pipelines.
- Washing pipes.
- Visiting the accessible parts of the pipe to control the sediment layer created on the surface of the inner wall of the pipes.

4.1.7 TANKS

4.1.7.1 PRE-LAUNCH OPERATION

Before starting the exploitation operation, it must be ensured that the following things are fully fulfilled:

- All problems, which caused the failure of the unit, are recorded in the operational sheets.
- The repair and replacement of the parts that were involved in the occurrence of problems has been done.
- The tank overflow system is ready for operation.
- The performance of the taps entering the tanks is guaranteed.
- The performance of the outlet taps from the tanks is guaranteed.
- The performance of the chemical injection and chlorination system is guaranteed.
- Preparing the daily report book (operational sheets).

Normal operation

- Continuous control of the height of the water level in the tank and record it in the relevant worksheet.
- Ensuring that the inlet and outlet valves are open.
- Ensuring the operation of ventilators and ventilation system.
- Inspection and notification of the absence of leaks in the walls and ceiling.
- Ensuring the operation of the inspection valves.
- Visiting the appearance of the water.
- Record the report of each of the above cases in information sheets.
- Continuous control of chemical injection rate.

4.1.7.2 MAINTENANCE AND LUBRICATION

The schedule of inspection and maintenance of tanks are as the following:[27,47–53]

Periodic visit schedule of reservoirs

- Row of operation period Description of operation.
- Quarterly inspection of internal and external surfaces.
- General inspection of faucets and ensuring their correct operation.
- Inspection of manholes and repair of possible damage to them.
- Inspecting the roof vents and making sure they are healthy.
- Taking the tank out of service every 2 years.
- Full inspection of internal and external surfaces in terms of leaks and possible damages.
- General services of faucets and necessary repairs.
- Washing and disinfecting the tank.
- Check existing valves.
- The outer coating of faucets and connections is painted to protect against corrosion.
- Inspection of the gearbox, the butterfly valves should be replaced and adjusted if necessary.

Repairs

- Examining faucets and other connections and repairing them if necessary.
- Tanks should be checked for leaks and if necessary, isolation should be done.
- Investigate the installation locations of valves and ventilators, and in the case of infiltration of surface water and runoff through the wall of the valves and ventilators, necessary repairs should be made.

Safety

- Preventing the entry of surface water, as a result of atmospheric spills that have accumulated on the tank as waste water.
- Full sealing of the inspection valves.
- Ensuring that all the outlet valves of the tank are left-handed.
- Preventing and hindering the entry of different people.

4.2 METHODS

The current working method includes two separated parts of investigation for invisible water leakage. First due to invisible leakage of water distribution network pipes and second for water transmission line. The selected method for invisible leakage finding was the correlation method. For water distribution network accident was investigated. For water transmission line, the relationship between water hammer impact and water transmission line breakage were investigated. Finally, a dynamic regression model was presented for the investigation of relationship between water hammer impact and the location of water leakage. This work is semi-experimental and the researcher considers to use the test of water transmission lines to determine the destructive position of the "water hammer impact" phenomenon related to invisible leakage.

4.2.1 VARIABLES OF REGRESSION MODEL

Variables related to the research questions are: wave speed, pressure caused by shock wave, water loss, leakage position.

There is a significant relationship between the wave speed and the pressure caused by the shock wave (4.5):

$$V = f(H) \tag{4.5}$$

There is a significant relationship between the pressure caused by the shock wave and water loss (4.6):

$$H = f(Q) \tag{4.6}$$

There is a significant relationship between water loss and wave speed and pressure caused by the shock wave (4.7)

$$Q = f(V, H) \tag{4.7}$$

There is a significant relationship between the location of the leak and the pressure caused by the shock wave (4.8):

$$S = f(H) \tag{4.8}$$

S: Position of leakage (m). Q: Water loss (L/s).H: Pressure caused by shock wave (bar).V: Wave speed (m/s).

The work was conducted by field method using SPSS13 software/Hammer7 software/Gambit software, Fluent software/Arc-GIS10 software.

While taking advantage of regression analysis and performing ANOVA and *t*-test, a dynamic model was defined.

4.2.2 REGRESSION ANALYSIS

The use of one variable to predict another variable is called regression. Regression analysis is a statistical technique for examining and modeling the relationship between variables. Regression analysis is used in almost any field, including engineering, physics, economics, management, Life sciences, biology and humanities, and social sciences for estimation and prediction. Regression analysis is one of the most widely used methods among statistical techniques. Regression, with the use of a specific known variable, predicts the values of another unspecified variable. The amount of change of one variable due to another variable is also called regression coefficient, which is the amount of change that occurs in the dependent variable as a result of one unit change in the independent variable. Regression is one-variable and two-variables are calculated. In one-variable regression, there is one independent variable and one function variable, but in two-variable regression, there is one function variable and two independent variables. To begin with, there must be a linear relationship, which is the Scatter Diagram. The basic idea of this gives the subject. The regression line reflects the general movement path of the scattered points in the nominal coordinate system, which can show the strength and weakness and the type of correlation between the variables. The regression equation should be used to draw the regression line.

4.2.3 RESEARCH TOOLS

These tools are included for data receiving from pressure gauges and flowmeters with remote reading loggers to measure wave speed, shock wave pressure, water loss, leak location, and record them. As soon as the pump stopped suddenly, the loggers first recorded the data received from the pressure gauges and flowmeters. In this regard, hydraulic instability data analyzer sensors functioned with ultra-advanced signal receiving technology in a fraction of a second and in accordance with the geographic information system (GIS) for the installation of water transmission lines. Recording the location and reference ground coordinates of the water transmission lines in the computer database under the management of the GIS, as well as the simultaneous information exchange of the receiver and transmitter equipment through the logical and programmable PLC control computer system, made it possible to determine the location of the leak and record it.

The measurement tools included the following:

- Cut maps in CAD and SCAN format.
- GPS.
- Logger flowmeters for remote reading.
- Remote reading pressure gauges.
- HAMMER software
- ArcGIS9-ArcMapsoftware.
- SPSS13 software
- Gambit software
- Fluent software

4.2.4 PART ONE: WATER DISTRIBUTION NETWORK PIPES

4.2.4.1 POPULATION AND STATISTICAL SAMPLE

The statistical population of this work includes the water distribution network pipes, which was investigated in order to reduce non-revenue water (NRW) and save water distribution.

In the new global view, UFW in the distribution network includes the following situations:

- Water losses due to fractures and incidents (visible leakage).
- Water losses due to small leaks (invisible leaks).
- Water losses from faucets and fittings.
- Removal of fire extinguishers.
- Washing the network through the available drain valves.

Small leaks caused by faucets and fittings are considered to be part of small or invisible leaks. In this way, the minimum amount of water not taken into account due to faucets and fittings can be calculated.

Since leaks caused by broken pipes and connections in the water distribution network are identified and fixed after receiving a leak report from local residents, leaks caused by network failures and incidents are considered visible types of leaks.

4.2.4.2 WATER IS NOT TAKEN INTO ACCOUNT DUE TO NETWORK ACCIDENTS AND BRANCHES

Unaccounted water due to network accidents and branches is part of visible leaks. The occurrence of breakdowns and accidents is generally caused by non-observance of design and implementation standards during the creation of distribution networks, as well as due to urban constructions and network development programs.

To determine the amount of unaccounted water caused by broken pipes in the distribution network, Bernoulli and continuity relations have been used.

The information used in these calculations is based on information from the worksheet of distribution network events, the average amount of broken pipes throughout the year, the results of the calibrated network analysis using pressure measurement results. For this purpose, the information obtained from the above is given in a table according to the table below (Table 4.23).

TABLE 4.23 Calculation of Unaccounted Water Caused by Failure and Events (Sample Form).

Water losses cm^2	Output flow m^3/h	Output water speed m/s	Average water pressure in the pipe m/s	Average speed of water in the pipe m/s	Pipe fracture surface cm^2	Pipe cross-section cm^2	Leak time		The number of fracture	Pipe diameter and material
							Day	The watch		

4.2.4.3 NUMBER OF FRACTURES

Determining the number of breaks for measuring the amount of unaccounted water caused by network breaks in two ways of fractures is done while completing the incident worksheet in each month.

Due to the fact that there is an average amount of consumption in the network in this month, the statistics of incidents related to this month can be considered an average of the number of fractures during the months of the year.

In the distribution network, during the year, there is a possibility of accidents that did not happen in the month of collecting the information related to the accidents by the consultant. Therefore, according to the statistics related to accidents in different years and the review and evaluation of this information, the average amount of these accidents according to the probability of their occurrence per month has been included in the calculations.

4.2.4.4 THE DURATION OF LEAKAGE

The duration of leakage due to breakages is the period of time between the moment the pipe breaks or bursts until the moment the leakage stops and the defect is fixed. This time is divided into smaller parts as described below.

4.2.4.5 THE LEAK WAS OBSERVED ON THE GROUND

It is the time it takes for the water to reach the surface of the ground from the place of the fracture, and the place of the fracture is identified as the accumulation of water on the surface of the ground. This time will vary according to the diameter of the pipe, the type of pipe, type of soil, the ground cover, and the level of underground water.

This time for the breakage of pipes with a small diameter is very long and even reaches a month, and the mentioned time decreases gradually as the diameter of the breakage increases. In this sense, one of the ruling engineering relationships has been used during this kind of relationship leakage.

In this research, the main focus is to fix this type of physical leaks. Big leaks are quickly identified on the ground level, it is people's reports or reports of periodical visits, which repair crews quickly take action to fix the leak.

4.2.4.6 INFORMING LOCAL RESIDENTS

The water leakage caused by the rupture and bursting of the pipe is revealed on the ground until the moment when the local people are informed about it. It takes a period of time, which is called the time of informing the local residents. The people of the locality find out about the failure when water collects in the alley or street, or the water is cut off in a part of the place, or they face a severe drop in water pressure. This time increases at night due to the reduction of traffic in the streets. There have been cases of water breaks in the streets at night due to the increase in network pressure, but with the beginning of the day and the decrease in pressure and evaporation of water, its location is not very noticeable, and thus, a long time has passed since the time of the break before people are aware of it and get informed about this.

Due to the weather, there is a possibility that the residents will not immediately report the accumulation of water on the surface of the streets.

4.2.4.7 INFORMATION FROM THE PEOPLE

Considering the vital importance of water and in order to prevent further water loss, it is expected that the local residents will inform at the first opportunity after identifying the location of the leakage. Therefore, by informing as many people as possible, we should try to shorten the average time of people contacting the company for information and not exceed 0.5 to 1 h.

4.2.4.8 REFER TO THE PLACE

It takes some time for the accident team to visit the place after being informed about the location of the failure. In order to estimate the average time to visit the place, the use of GIS software is considered to determine the location of the fracture site.

In order to reduce the time to go to the place, it is necessary to study the effects of equipping the relevant personnel with wireless according to the external conditions of each city.

4.2.4.9 FINDING THE LOCATION OF THE FAILURE

It is the time that it takes to identify the location of the break and start digging the ground and determine the location of the leak on the pipe.

4.2.4.10 RECLAMATION OF FAILURE

It is the period of time when, after finding a break, it is possible to temporarily or permanently stop the water leakage with the available facilities.

4.2.4.11 THE PROBABILITY OF BREAKAGE AT NIGHT OR DAY

Breaks usually happen at night with the increase of pressure in the network, therefore, according to the results of the incident worksheets, the percentage of breaks is determined during day and night hours. Due to the difference in the duration of the fractures during the day and night, the above percentages will be included as a factor in the average time.

It has been determined with the investigations that more than 60% of the fractures are related to night fractures and 40% are related to daytime fractures

According to the above, the minimum time required to repair each broken item in the distribution system is given in Table 4.24.

TABLE 4.24 The Minimum Time Required to Repair Each Broken Item.

Total time after detection	Temporarily cutting off water or repairing a break	Identifying the location and digging the ground	Refer to the place	Contact the company for information	Inform the local residents	Time to reveal	Distribution of the average time of leakage from the fracture
(h) 9	(h) 0.5	(h) 0.5	(h) 1	(h) 1	(h) 6	–	The night
(h) 5/3	(h) 0.5	(h) 0.5	(h) 1	(h) 0.5	(h) 1	–	Day

According to what was mentioned, it is possible to calculate the average duration of leakage by separating the type and diameter of the pipes in the calculation of the water not taken into account due to the fractures as follows:

$$(Hours)\ 9 * 60\% + 5.3 * 40\% = 8.6, \tag{4.9}$$

This time is related to the continuation of the leak after informing the local residents until the water is cut off and the leak is fixed, and does not include the time it was discovered.

Leak detection equipment: using leak detection equipment such as advanced leak detection device, allows the leaks in the network to be identified before they are detected on the ground (Figures 4.6 and 4.7).

The leakage location detection by correlation analysis and leakage location detection by ultrasonic flowmeter (UFM) (Figures 4.8 and 4.9) are given in the following equations (4.10–4.11):

$$d1 + d2 = D \tag{4.10}$$

where D: total length of the pipeline (distance between sensor 1 and sensor 2). d1: leakage location distance to sensor 1. d2: leakage location distance to sensor 2.

FIGURE 4.6 Identified leakage on the ground.

FIGURE 4.7 Identified leakage on the water pipeline

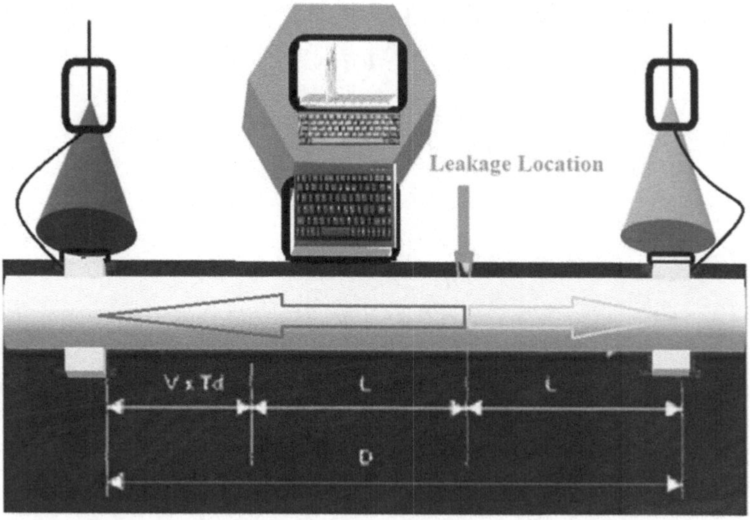

FIGURE 4.8 Leakage location detection by correlation techniques.

FIGURE 4.9 Leakage location detection by ultrasonic flowmeter (UFM).

$$V = D/t, \tag{4.11}$$

where *v*: speed of the correlator pulse wave.*t*: detection time for correlator pulse wave travel from leakage location to sensor.

t1 and t2 can be calculated as (4.12–4.21):

$$d1 = Vt1 \tag{4.12}$$

$$d2 = Vt2 \tag{4.13}$$

$$d2 - d1 = Vt2 - Vt1 \tag{4.14}$$

$$d2 - d1 = (t2 - t1) \tag{4.15}$$

$$d2 - d1 = V\Delta t \tag{4.16}$$

$$d1 + d2 = D \tag{4.17}$$

$$d2 - d1 = V\Delta t \tag{4.18}$$

$$2d2 = D + V\Delta t \tag{4.19}$$

$$d2 = D/2 + V\Delta t/2 \tag{4.20}$$

$$d1 = D - \frac{D}{2} + V\Delta t / 2) \tag{4.21}$$

This reduces the amount of water loss. Also, network leaks that cannot occur on the ground are identified and repaired by using GIS maps.

The distribution network and maps of the accident site will reduce the duration of leakage due to broken pipes. The duration of the leakage for the broken pipes with small diameter is very long and even reaches a month, and gradually the said time decreases as the diameter of the pipe becomes bigger and broken. In this research, the main focus is on fixing this type of fractures and leaks.

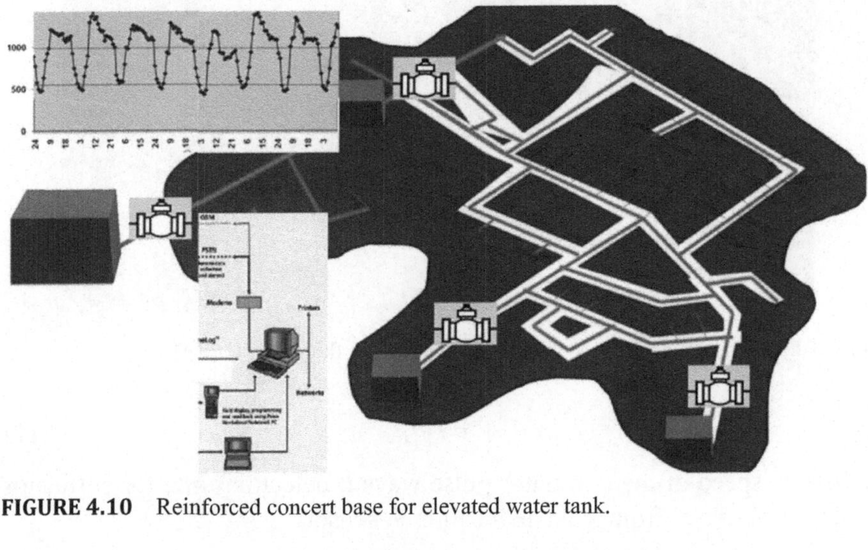

FIGURE 4.10 Reinforced concert base for elevated water tank.

The detection time of small leaks is usually very long. According to the correct principles and standards, including soil permeability coefficient, the minimum detection time can be calculated in pipes with a small diameter (less than 3 inches).

According to the general characteristics of the soil, the permeability coefficient of the mentioned soil can be considered.

The distribution network pipes are usually located at a depth of 1.20 meters from the ground surface.

This point is important that in this way, the minimum amount of casualties caused by breakages and accidents is calculated.

TABLE 4.25 The Minimum Average Time for the Continuation of the Leak, Which Is Considered in the Calculations.

Duration of leakage (days)	Pipe diameter (mm)
14	12.5–25
14	45–50
7	75–100
3	150–200
1.5	≥250

4.2.4.12 FRACTURE CROSS-SECTION

In order to determine the cross-section of the pipelines, the information obtained from the incident worksheet was used. For the cases where this information was not available, the cross-sectional area of the fracture or pores caused by corrosion was considered to be 5% of the total cross-sectional area of each pipe (Figures 4.11 and 4.12).

4.2.4.13 AVERAGE SPEED OF WATER IN THE PIPE

In this work, by using the results of the calibrated network analysis, first the flow velocity in each pipe diameter was calculated and then the average velocity in different diameters was determined. This amount is close to each other in different diameters and is considered the same for all diameters.

FIGURE 4.11 Cross-sectional area of underground pipes; different locations.

FIGURE 4.12 Cross-sectional area of underground pipes; collected in main channel.

4.2.4.14 AVERAGE WATER PRESSURE

In order to determine the average water pressure in different points of the distribution network, the results of pressure gauge and calibrated hydraulic model were used. For each region of the pilot range, the average pressure during the day and night was taken into account.

4.2.4.15 PRESSURE REDUCTION CAUSED BY LEAKAGE

In any case, water leakage from the pipe and discharge from the pipe increased. The pressure inside the pipe was decreased or fallen. Based on the calculation bases and standards and local measurements, a percentage of the decrease in the pressure inside the pipe after the fracture compared with the pressure in the normal state (about 50%) has been estimated.

4.2.4.16 FORMULATION

The method of calculating NRW due to fractures:

In order to determine the amount of NRW due to pipe breaks in the distribution network, continuity and Bernoulli relations are used (4.22–4.28).

Bernoulli relation

$$gz + \frac{V^2}{2} + \frac{P}{\rho} = Cte \tag{4.22}$$

where: g: acceleration of gravity (m/s²); Z: vertical height (m); v: fluid velocity (m/s); p: fluid pressure (pa); ρ: density (kg/m³).

Because the right side of both relations is equal to a constant value, by making the left side of the written equations equal and replacing the collected information, some parameters are removed from both sides and the equation in terms of V 2 is obtained as follows:

$$(gz1) + \frac{V1^2}{2} + \frac{P1}{\rho} = (gz2) + \frac{V2^2}{2} + \frac{P2}{\rho} \tag{4.23}$$

P2 = P0 (pa): pressure outside the pipeline (atmospheric pressure)P1: pressure inside the pipeline (pa) V1: water speed inside the pipe (m/s) V2: water speed while leaving the fracture site.

Due to the lack of height difference between water break levels (points 1 and 2) gz1 and gz2 are removed from the sides of the equation and the equation is arranged in terms of V2 as follows:

$$(gz1)+\frac{V1^2}{2}+\frac{P1}{\rho}=(gz2)+\frac{V2^2}{2}+\frac{P2}{\rho} \quad (4.24)$$

$$(gz1)+\frac{V1^2}{2}+\frac{P1}{\rho}=(gz2)+\frac{V2^2}{2}+\frac{P2}{\rho} \quad (4.25)$$

$$(gz1)+\frac{V1^2}{2}+\frac{P1}{\rho}=(gz2)+\frac{V2^2}{2}+\frac{P2}{\rho} \quad (4.26)$$

The speed of water when leaving the fracture site

$$(gz1)+\frac{V1^2}{2}+\frac{P1}{\rho}=(gz2)+\frac{V2^2}{2}+\frac{P2}{\rho} \quad (4.27)$$

Using the continuity relationship and knowing the speed and cross-sectional area, the output water amount (water loss) is calculated.

Continuity relationship

$$Q = V2 * A \quad (4.28)$$

Q: volume of water past the known cross-section (m^3/s);A: fracture cross-section (m^2);V2: water speed when leaving the fracture section (m/s)

The calculations related to the losses caused by accidents and fractures have also been done with the theoretical AWWA method, and the comparison of checking and performing leakage calculations with both methods gives almost the same results.

NRW due to invisible leakage

In this work, due to the fact that the faucets of the distribution network are underground, it was not possible to calculate small leaks from the faucets through visual inspection. These cases were considered as small leaks in the calculations. In this way, the minimum amount of NRW caused by small leaks can be calculated. Large leaks that happen as an accident in the network are called breaks against small and localized leaks in the network that are not visible on the ground. Leaked water is probably mistaken for runoff water. Due to the evaporation of water from small leaks, its location is not very noticeable. We need systems of periodic visits in the network and objective observations and completion of the identity card of faucets and use. A leak detector can be used to fix invisible leakage. By using leak detection devices, such leaks can be identified.

Estimating the amount of leakage allowed in water distribution network lines

The amount of permissible leakage from pipes and connections has been determined by the following formula (4.29):

$$L = \frac{ND\sqrt{P}}{3.3} \tag{4.29}$$

L: allowable amount of leakage (cm^3/h); N: number of joints in the pipeline; D: pipe diameter (cm); P: average pressure inside the pipe (kg/cm^2)

By using the above relationship and determining the number of network inlet connections based on the length of the pipe unit, the amount of leakage has been calculated.

4.2.5 PART TWO: WATER DISTRIBUTION NETWORK PIPES

One of the most important factors for water transmission failure is happened by air entrance or air penetration due to water hammer into the water pipeline. By Geography Information Systems (GIS), this phenomenon can be recognized during irregular operations.

In this work, numerical analysis and simulation processes were applied for water hammer. With free gas systems and the potential for water-column separation, the numerical simulation of hydraulic transients was more complex. Small pressure spikes caused by the type of tiny vapor pockets that was difficult to simulate accurately seldom result in a significant change to the transient envelopes. During transient flow, larger vapor-pocket collapse events resulting in significant upsurge pressures were simulated with enough accuracy to support definitive conclusions. This work evaluated the invisible leakage of water pipeline by computational method such as finite differences (FD) method (4.30–4.39) for turbulent flows based on the GIS.

Computational methods for turbulent flows:

- direct numerical simulation (DNS)
- large Eddy simulation (LES)
- Reynolds-averaged Navier–Stokes (RANS)

Navier–Stokes equations:

$$((\partial t)+div(\rho\vec{u}))=0(Continuty\,equation) \tag{4.30}$$

$$(\partial t)+div(\rho u\vec{u}) = -(\partial x)+div(\mu gradu)+q_{x,}X(momentum\,equation) \tag{4.31}$$

$$(\partial t)+div(\rho v\vec{u}) = -(\partial y)+div(\mu gradv)+q_{y,}Y(momentum\,equation) \tag{4.32}$$

$$(\partial t)+div(\rho w\vec{u}) = -(\partial z)+div(\mu gradw)+q_{z,}Z(momentum\,equation) \tag{4.33}$$

$$(\partial t)+div(\rho i\vec{u}) = -pdiv\vec{u}+div(kgradT)+\varnothing,Z(Internal\,energy\,equation) \tag{4.34}$$

Direct numerical simulation (DNS) (4.6–4.8):

- Discretize Navier–Stokes equation on a sufficiently fine grid for resolving all motions occurring in turbulent flow.
- No uses of any models.
- Equivalent to laboratory experiment.
- Relationship between length η of smallest eddies and the length L of largest eddies.

$$L \approx \left(R_{e_L}\right)^{\frac{3}{4}} \tag{4.35}$$

Number of elements necessary to discretize the flow field

$$n_{elem} > 10^{13} \tag{4.36}$$

In industrial applications:

$$Re > 10^{6} \tag{4.37}$$

Large Eddy Simulation (LES)

- Only large eddies are computed.
- Small eddies are modeled with the sub-grid-scale (SGS) models.
- Reynolds-Averaged Navier–Stokes (RANS) (4.9–4.10)
- Variables decomposed in to a mean part and a fluctuating part.
- Navier-Stokes equations averaged over time
- Turbulence models are necessary.

$$u=\overline{u}+u' \tag{4.38}$$

$$H_2 - H_1 = (g)(v_2 - v_1) = (\rho C)(v_2 - v_1), \tag{4.39}$$

(*Joukowski formula*)

First of all, by remote sensing facility equipped with data loggers, the leakage points in the water pipeline were investigated. In order to reduce the real case amount of NRW, the relationship classes between the spatial data and non-spatial data as a tool for communication between coordination of water system elements related to NRW GEO-database were defined. Therefore, the relationship class for water system elements according to the type of connection (simple or complex) was provided and accessed to the leakage points. The hydraulic model of NRW for analysis of water loss was defined in compliance with geography information system (GIS) and by HAMMER, WATERGEMS, and ArcGIS9-ArcMap software.

4.3 RESULTS AND DISCUSSION

Based on the main results of this work, for invisible water leakage, there was a significant relation between the density of accident and percent of pipe breaks.

4.3.1 PART ONE RESULTS; WATER DISTRIBUTION NETWORK PIPES

Analysis of the number of incidents on network components:

- Based on the recorded statistics of the occurrence of accidents, an analysis was done to separate the areas of the distribution network, the results. Accidents on main lines accounted for 4.8%, subsidiary lines for 15.1%, and domestic branches accounted for 80.1% of accidents. Of the accidents that occurred on the branches, 43% are related to pipe bursts and 57% are related to leakage from the location of the meter or accidents on the meters.

Classification of incidents and events were as the following items:

- *Incident:* Many of the incidents that happened in the water distribution network are local and affect a small area of the distribution network for consumers and can be repaired in less than 24 h.

- *Incident:* It includes cases of water cut or shortage caused by defects in the main facilities, such as pumping stations, transmission lines, which have caused problems in a large part of the network and require more than 24 h to repair it.
- *Defects of branches:* Every day, in the urban water distribution network, the special branches of the subscribers suffer from many defects, the problem of which is unique to the particular subscriber and can be repaired quickly. Such incidents are very many and include most of the network defects.

4.3.2 EXAMINING THE STATISTICAL INDICATORS OF ACCIDENTS

In order to compare the status of incidents and events in the network, the following indicators have been presented:

- Index 1: Spatial abundance
- Index 2: Incidents per unit of network length
- Index 3: Defects of branches per unit of network length
- Index 4: Main incidents per unit of network length

The spatial frequency index is the indicator of the vulnerable area. A large number of incidents on the network and branches of this index generally indicates the number of incidents on the network and branches. Usually, they include more than 70% of all incidents in branches. The index of accidents per unit of length is on the main and secondary routes, except for the special routes of the branches. Such routes include pipes, valve connections, pumping stations and tanks.

4.3.3 PART TWO RESULTS; WATER PIPELINES

4.3.3.1 RELATION BETWEEN SHOCK WAVE PRESSURE AND WATER LOSS

Based on the main results of this work, for invisible water leakage, there was a significant relation between the pressure caused by the shock wave and water loss. The significance level of this hypothesis was slightly far from significance ($P = 0.065$). The possible reason for the non-significance of the relation between shock wave and water loss in this study can be attributed to the dispersion of water loss.

In any case, the shock wave and water loss are among the important indicators of water leakage detection, and design engineers should pay attention to it in the calculations.

4.3.3.2 RELATION BETWEEN WATER LOSS AND WAVE SPEED AND SHOCK WAVE PRESSURE

The results of this study showed that there was a significant relationship between water loss and wave speed and the pressure caused by the shock wave.

4.3.3.3 RELATIONSHIP BETWEEN THE LEAKAGE POSITION AND THE PRESSURE CAUSED BY THE SHOCK WAVE

The results of this work also showed that there is a significant relationship between the location of the leak and the pressure caused by the shock wave, the pressure caused by the shock wave was pointed out as an indicator of leakage that can have an effect on water loss.

HAMMER and ArcGIS9-ArcMap10 software carried out management of water leakage based on the production data and analysis of transmission process. In other words, "GIS" is a computer system for managing and analyzing the spatial information of leakage and the volume of water loss in the form of descriptive information, and it has the ability to collect, store, analyze and display spatial and non-spatial (descriptive) information at the same time.

Of course, this issue may be better manifested in the form of technological progress of receiving input data to the georeference model (in a fraction of a second) to determine the location of the leak, which certainly requires more research in this field.

Two-variable regression model (4.40–4.55) (Figures 4.13–4.21) (Tables 4.26–4.35)**:**

$$Q = f(V, H) \tag{4.40}$$

MODEL: MOD_1.

Independent: VAR00002 Transient Head MAX. (m-h)

Independent: VAR00003 Wave Speed (m/s)

Dependent Variable: Flow (L/s)

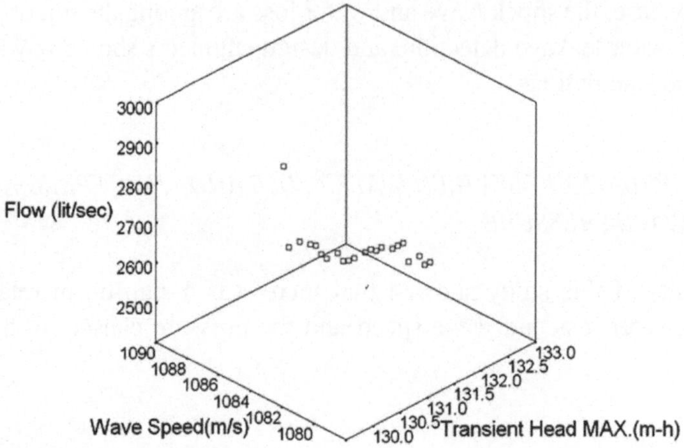

FIGURE 4.13 Two-variable regression model; scatter diagram.

TABLE 4.26 Model Summary.

Model	R	R square	Adjusted R square	Std. error of the estimate
1	.883[a]	.780	.769	44.34911

[a]Predictors: (Constant), Transient Head MAX (m-h).

TABLE 4.27 ANOVA[b]

Model		Sum of Squares	df	Mean Square	F	Sig.
1	Regression	139527	1	139527.718	70.94	.000[a]
	Residual	39336	20	1966.844		
	Total	178864	21			

[a]Predictors: (Constant), Transient Head MAX (m-h).

[b]Dependent Variable: Flow (L/s).

TABLE 4.28 Coefficients[a]

Model		Unstandardized coefficients		Standardized coefficients	t	Sig.
		B	Std. Error	Beta		
1	Constant	15560	1536.8		10.125	.000
	Transient Head MAX. (m-h)	−98.441	11.688	−.883	−8.423	.000

[a]Dependent Variable: Flow (L/s).

Independent: VAR00002 Transient Head MAX (m-h).

Independent: VAR00003 Wave Speed (m/s).

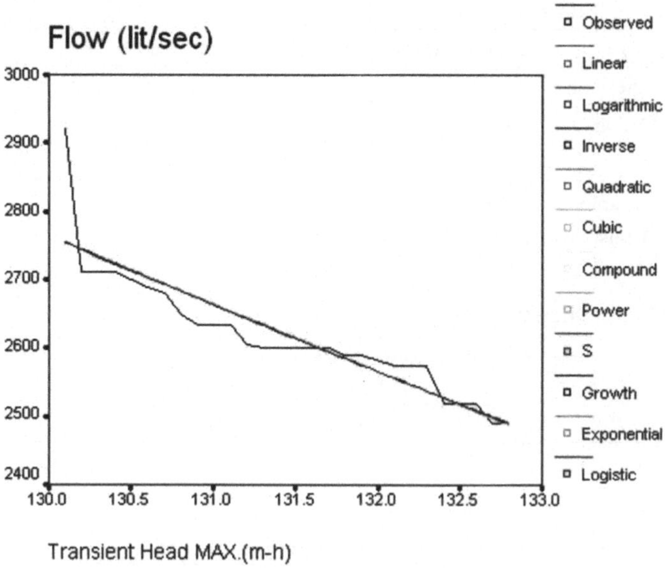

FIGURE 4.14 Regression model (Flow–Transient Head); curve fit.

One-variable regression model

$$S = f(H) \tag{4.41}$$

MODEL: MOD_2.

Independent: VAR00002 Transient Head MAX (m-h)

Dependent Variable: Distance (m)

TABLE 4.29 One-Variable Regression.

Model	Variables entered	Variables removed	Method
1	Transient Head MAX (m-h)[a]	.	Enter

[a]All requested variables entered.

Dependent variable: Distance (m).

TABLE 4.30 Model Summary.

Model	R	R square	Adjusted R square	Std. error of the estimate
1	.968[a]	.937	.934	24.33516

[a]Predictors: (Constant), Transient Head MAX (m-h).

TABLE 4.31 ANOVA[a].

Model		Sum of squares	df	Mean square	F	Sig.
1	Regression	176126	1	176126.197	297.410	.000[b]
	Residual	11844	20	592.200		
	Total	187970	21			

[a]Dependent variable: Distance (m).

[b]Predictors: (Constant), Transient Head MAX (m-h).

TABLE 4.32 Coefficients[a].

Model		Unstandardized coefficients		Standardized coefficients	t	Sig.
		B	Std. error	Beta		
1	(Constant)	14700	843.3		17.433	.000
	Transient Head MAX. (m-h)	−110	6.413	−.968	−17.246	.000

[a]Dependent Variable: Distance (m).

Independent: VAR00002.

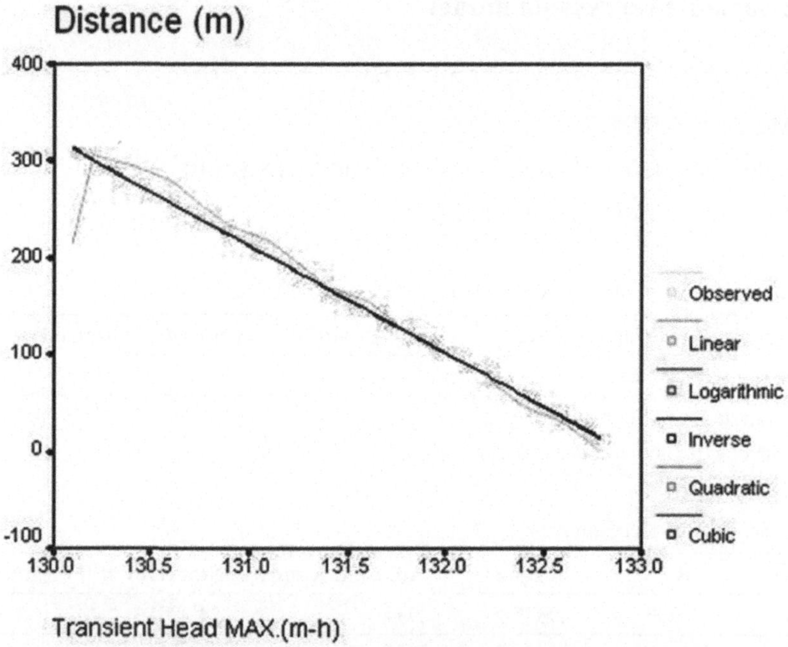

FIGURE 4.15 Regression model (Distance–Transient Head); curve fit.

One-variable regression model

$$H = f(Q) \tag{4.42}$$

MODEL: MOD_3.
Dependent Variable: Transient Head MAX (m-h).
Independent: VAR00002 Flow (L/s).

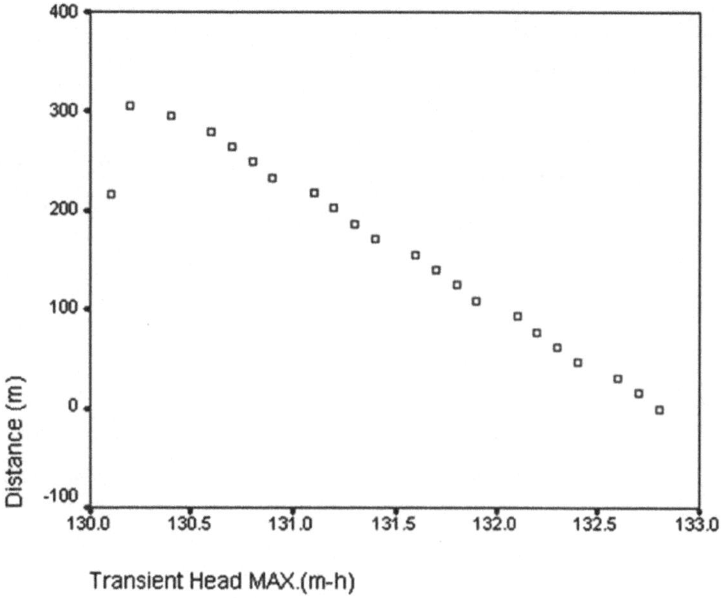

FIGURE 4.16 Scatter diagram (Distance–Transient Head).

TABLE 4.33 Model Summary.

Model	R	R square	Adjusted R square	Std. error of the estimate
1	.883[a]	.780	.769	.39790

[a]Predictors: (Constant), Flow (L/s).

TABLE 4.34 ANOVA[a].

Model		Sum of squares	df	Mean square	F	Sig.
1	Regression	11.232	1	11.232	70.940	.000[b]
	Residual	3.167	20	.158		
	Total	14.398	21			

[a]Dependent Variable: Transient Head MAX (m-h).
[b]Predictors: (Constant), Flow (L/s).

TABLE 4.35 Coefficients[a].

Model		Unstandardized coefficients		Standardized coefficients	t	Sig.
		B	Std. error	Beta		
1	(Constant)	152.222	2.463		61.808	.000
	Flow (L/s)	−.008	.001	−.883	−8.423	.000

[a]Dependent Variable: Transient Head MAX (m-h).

Dependent Variable: Transient Head MAX (m-h)

Independent: VAR00002 Flow (L/s).

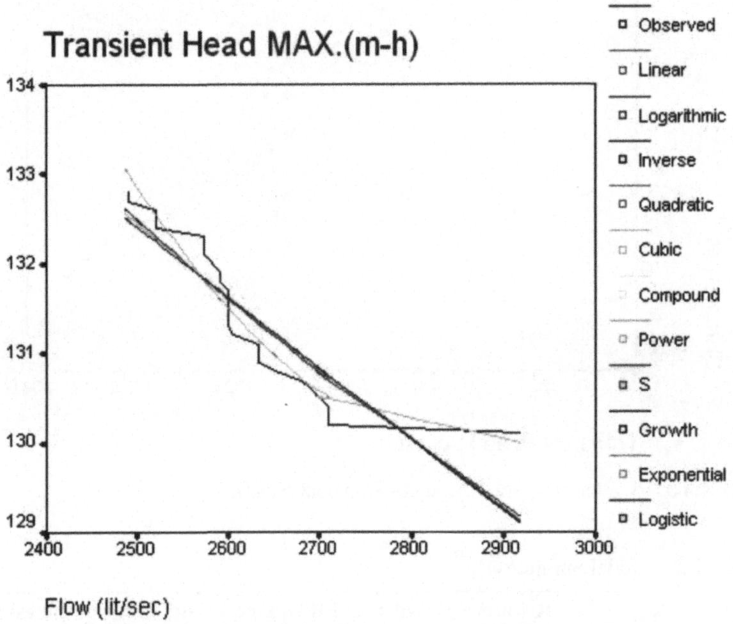

FIGURE 4.17 Regression model (Flow–Transient Head); curve fit.

One-variable regression model

$$V = f(H) \tag{4.43}$$

MODEL: MOD_4.

Dependent Variable: Wave Speed (m/s).

Independent: Transient Head MAX (m-h).

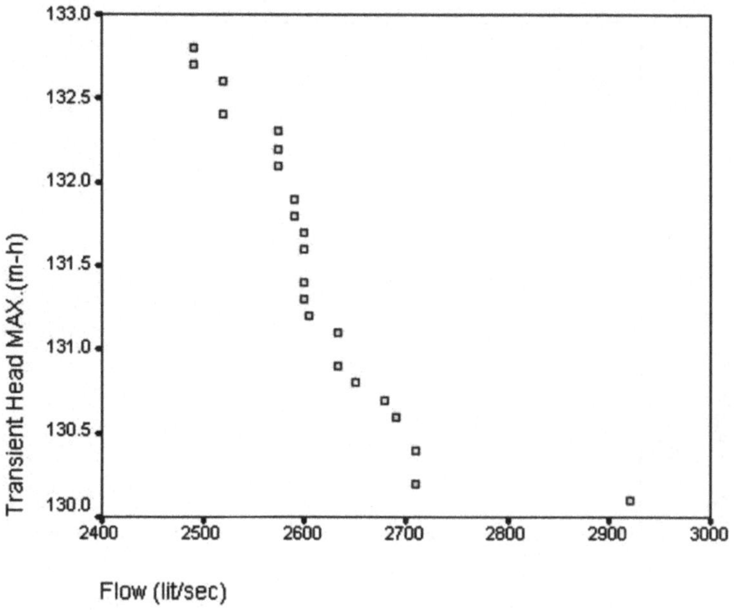

FIGURE 4.18 Scatter diagram (Flow–Transient Head).

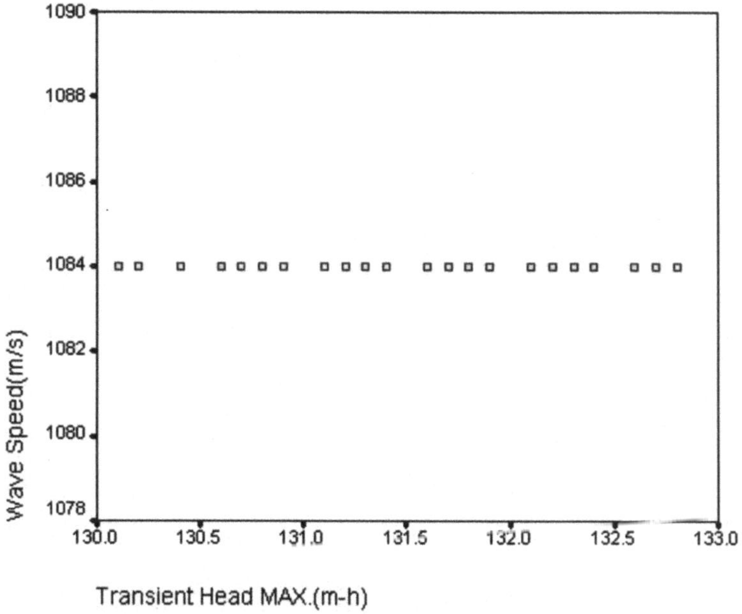

FIGURE 4.19 Scatter diagram (Wave Speed–Transient Head).

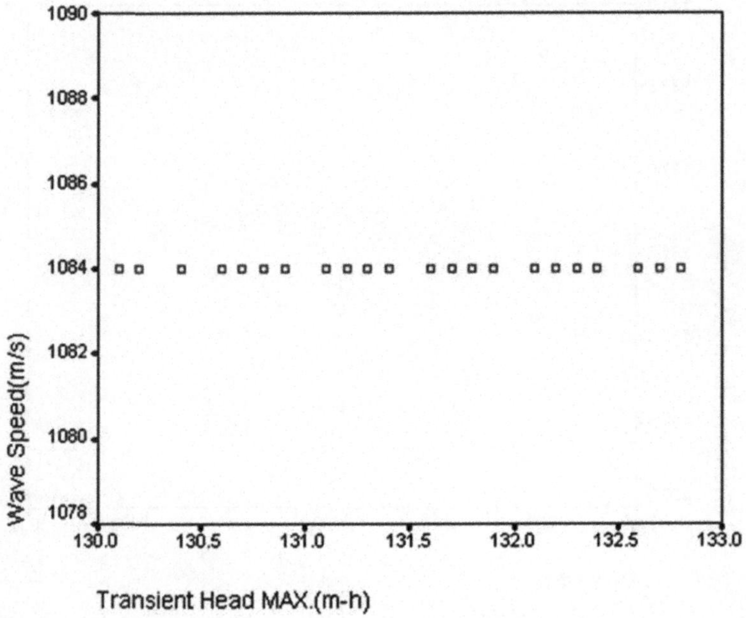

FIGURE 4.20 Scatter diagram (Wave Speed–Transient Head).

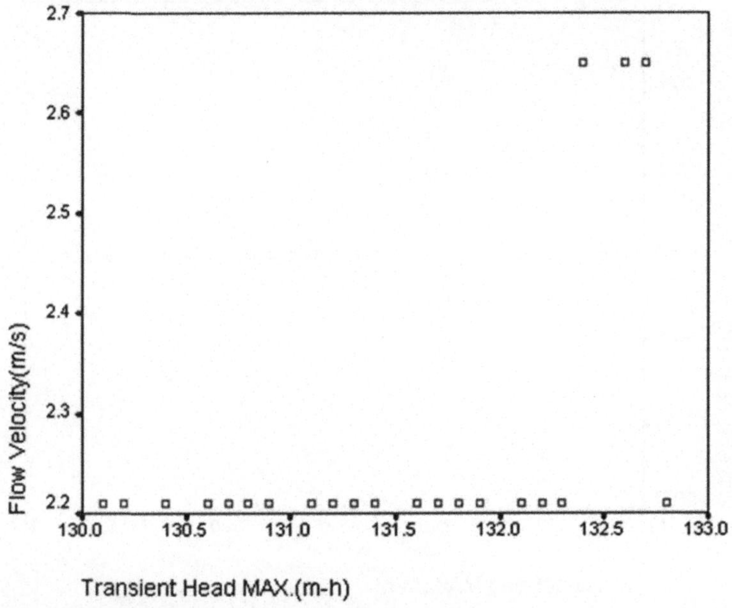

FIGURE 4.21 Scatter diagram (Flow–Transient Head).

One- and two-variables regression model

In the field method, by using the regression model, the data received from pressure gauges and flowmeters with remote reading for wave speed. The pressure caused by shock wave, water loss, leakage position will be measured and recorded. The curve of the estimated method is identified using these data, which is the regression test. These data were collected and analyzed from the real and existing system. This model was calibrated using a set of data without changing the parameter values. Formulas were written by using the obtained coefficients for the equation (4.44–4.55).

V = f (H), H =f (Q), S = f (H), Q = f (V, H).

$$y = a_0 + a_1 x \quad \text{(Line Equations)} \quad (4.44)$$

$$y = a_0 + a_1 x + a_2 x^2 \quad \text{(Second-order Equation)} \quad (4.45)$$

$$y = a_0 + a_1 x + a_2 x^2 + a_3 x^2 \quad \text{(Third-order Equation)} \quad (4.46)$$

$$y = a_0 + a_1 x + a_2 x^2 + a_3 x^3 + a_4 x^3 \quad \text{(Forth-order Equation)} \quad (4.47)$$

$$y = a_0 + a_1 x + a_2 x^2 + \ldots + a_n x^n \quad \text{(n-th order Equation)} \quad (4.48)$$

$$y = 1/(a_0 + a_1 x) \text{ or } 1/y = (a_1 x + a_0) \quad \text{(Hyperbola Equation)} \quad (4.49)$$

$$y = ab^x \text{ or } \log a + x \log b = a_0 + a_1 x \quad \text{(Exponential Equation)} \quad (4.50)$$

$$y = ax^b \text{ or } \log y = \log a + b \log x \quad \text{(Logarithmic Equation)} \quad (4.51)$$

$$y = ab^x + g \quad \text{(Expression Exponential Equation)} \quad (4.52)$$

$$y = ax^b + g \quad \text{(Expression Logarithmic Equation)} \quad (4.53)$$

$$y = pqb^x \;\&\; \log y = \log p + b^x \log q = ab^x + g \quad \text{(Complex Equation)} \quad (4.54)$$

$$y = pq^{bx} + h \quad \text{(Expression Complex Equation)} \quad (4.55)$$

4.4 CONCLUSION

In general, taking into account the findings of the present study, it is concluded that there is probably an indirect relationship between the leakage parameters and the volume of water wastage, as well as a direct relationship between the spatial and descriptive information with the subject of this research. Therefore, it is possible to advise the experts involved in the design, implementation, and maintenance of water transmission lines to use intelligent pressure management in the context of GIS to prevent the possibility of leakage and reduce the volume of water wastage.

4.4.1 PRACTICAL SUGGESTIONS ARISING FROM RESEARCH

In the process of research and analysis of the hydraulic model of the actual conditions, the following practical suggestions were considered:

- Creating a smart system of water transmission line by introducing tolls.
- Creating traceability and performing hydraulic analysis on the line of water transmission facilities.
- Transmission line modeling in ARCGIS system based on AMR remote reading data to control flow rate and pressure and the incident caused by RAM impact.

4.4.2 SUGGESTIONS FOR FUTURE RESEARCH

Relationship classes between spatial and non-spatial data are a tool to create a spatial relationship between complications and waste-related data. Therefore, it is suggested that for future research, relationship classes for the desired complications according to the type of relationship (simple or composite) and its amount has been determined and with this method, access to leakage points can be provided. Therefore, in brief, the title of the proposal for future research is as follows:

Determining leakage locations of the water transmission line using the definition of relationship classes for investigation of invisible leakage.

KEYWORDS

- **invisible water leakage**
- **Internet of things**
- **remote sensing**
- **distribution network**
- **networked sensors**

REFERENCES

1. Hariri Asli, H.; Aliyev, S. A. O.; Thomas S.; Deepu A. *Handbook of Research for Fluid and Solid Mechanics: Theory, Simulation, and Experiment*; Apple Academic Press, Taylor & Francis Group: New York. https://doi.org/10.1201/9781315365701.

2. Asli, H. H.; Hariri Asli, H.; Arabani, M.; Golpour, Y. Reclaimed Asphalt Pavement (RAP) Based on a Geospatial Information System (GIS). *Slovak J. Civ. Eng. Slovak Univ. Technol. Bratislava, Slovak* **2020,** *28* (2), 36–42. https://doi.org/10.2478/sjce-2020-0013.
3. NZS 3106, Design of Concrete Structures for the Storage of Liquids, Cement and Concrete Association of New Zealand, Standards Association of New Zealand, 2009.
4. Recommended Standards for Water Works, Great Lakes—Upper Mississippi River Board of State and Provincial Public Health and Environmental Managers, Health Research Inc.; Health Education Services Division, Albany, 2012.
5. Planning and Infrastructure, Above Ground Circular Concrete Tank, Manager Engineering, 2008. 103-Selected Earthquake Engineering Papers of George W. Housner, American Society of Civil Engineers, 1990.
6. Specification for Construction of Concrete Reservoirs, Standard Specifications and Drawings Specification for Construction of Concrete Reservoirs, Gold Coast, Australia, 2004.
7. TM 5-809-10. *Seismic Design Guidelines for Essential Buildings*; US Department of the ARMY, Washington, DC, 1986.
8. USEPA Finished Water Storage Facilities, U.S. Environmental Protection Agency, 2002.
9. Zecchin, A. C.; Maier, H. R.; Simpson, A. R.; Leonard, M.; Nixon, J. B. Parametric Study for an Ant Algorithm Applied to Water Distribution System Optimization. *IEEE Transitions Evol. Comput.* **2005,** *9* (2), 175–191.
10. Walski, T. M. Hydraulic Design of Water Distribution Storage Tanks. In: *Water Distribution Systems Handbook*; Mays, L. W., Ed.; McGraw-Hill: New York, 2000; Chapter 10.
11. Farmani, R.; Walters, G.; Savic, D. Evolutionary Multi-Objective Optimization of the Design and Operation of Water Distribution Network: Total Cost vs. Reliability vs. Water Quality. *J. Hydroinf.* **2006,** *8* (3), 165–179.
12. Haarhoff, J.; Van Zyl, J. E.; Nel, D.; Van der Walt, J. J. Water Supply and Water Quality in Rural Areas: Stochastic Analysis of Rural Water Supply Systems. *Water Supply* **2000,** *18* (1), 467–470.
13. Asli, H. H.; Investigation of the Factors Affecting Pedestrian Accidents in Urban Roundabouts. *Trans. Civ. Environ. Eng.* **2022,** *8*, Article ID: 2226, 1–4, Special Issue: NCTT 2021. https://doi.org/10.52547/crpase.8.1.2255.
14. Asli, H. H.; M. Arabani, Analysis of Strain and Failure of Asphalt Pavement. *Comput. Res. Progress Appl. Sci. Eng. Trans. Civ. Environ. Eng.* **2022,** *8*, 1–11, Article ID: 2250. https://doi.org/10.52547/crpase.8.1.2250.
15. https://en.wikipedia.org/wiki/Computational_physics
16. Maier, H. R.; Simpson, A. R.; Zecchin, A. C.; Foong, W. K.; Phang, K. Y.; Seah, H. Y.; Tan, C. L.Ant Colony Optimization for the Design of Water Distribution Systems. In: *Proceedings of World Water and Environmental Resources Congress 2001*; ASCE, The Rosen Plaza Hotel, Orlando, 2001.
17. Maier, H. R.; Simpson, A. R.; Zecchin, A. C.; Foong, W. K.; Phang, K. Y.; Seah, H. Y.; Tan, C. L. Ant Colony Optimization for Design of Water Distribution Systems. *J. Water Resour. Plann. Manag.* **2003,** *129* (3), 200–209.
18. Walski, T. M.; Chase, D. V.; Savic, D. A.; Grayman, W.; Beckwith, S.; Koelle, E. *Advanced Water Distribution Modeling and Management*; Haestad Press: Watertown, 2003.
19. Hashemi, S. S. *Optimization of Water Networks with Minimization of Pumping Energy*. Master Thesis, University of Tehran, Iran, 2010.

20. Mau, R. E.; Boulos, P. F.; Clark, R. M.; Grayman, W. M.; Tekippe, R. J.; Trusell, R. R. Explicit Mathematical Models of Distribution Storage Water Quality. *ASCE J. Hydraul. Eng.* **1995,** *121* (10), 669–709.
21. Rossmann, L. A. EPANET 2 User's Manual. Water Supply and Water Resources Division. U. S. Environmental Protection Agency. Cincinnati, EE.UU, 2000. http://www.epa.gov/ORD/NRMRL/wswrd/epanet.html
22. Asli, H. H.; Nazari, S. Water Age and Leakage in Reservoirs: Some Computational Aspects and Practical Hints. *Larhyss J.* **2021,** *48*, 151–167. http://larhyss.net/ojs/index.php/larhyss/article/view/808/807.
23. Yeung, H. Modeling of Service Reservoirs. *J. Hydroinf.* **2001,** *3* (3), 165–172.
24. Mahmood, F.; Pimblett, J. G.; Grace, N. O.; Grayman, W. M. Evaluation of Water Mixing Characteristics in Distribution System Storage Tanks. *JAWWA* **2005,** *97* (3), 74–88.
25. Asli, H. H.; Hozori, A. Non-Revenue Water (NRW) and 3d Hierarchical Model for Landslide. *Larhyss J.* **2021,** *48*, 189–210. http://larhyss.net/ojs/index.php/larhyss/article/view/810/810.
26. Marek, M.; Stoesser, T.; Roberts, P. J. W.; Weitbrecht, V.; Jirka, G. H. CFD Modeling of Turbulent Jet Mixing in a Water Storage Tank. In: *XXXII IAHR Congress*, Venice (Italy), July 1/6, 2007.
27. Van Zyl, J. E.; Haarhoff, J. Reliability Analysis of Municipal Storage Reservoirs Using Stochastic Analysis. *J. South African Inst. Civ. Eng.* **2007,** *49* (3), 27–32.
28. Asli, H. H.; Khodaparast Haghi, A.; Hariri Asli, H.; Sabermaash Eshghi, E. *Water Hammer Modelling and Simulation by GIS, Modelling and Simulation in Engineering*, Vol. 1, No. 3; Hindawi Publishing Corporation, 2012; Article ID 704163, 4 pages. DOI: 10.1155/2012/704163. http://www.hindawi.com/journals/mse/aip/704163/.
29. Manning, G. P. *Concrete Reservoirs and Tanks*; Concrete Publication, 1972.
30. *Minimum Design Loads for Buildings and Other Structures*; American Society of Civil Engineers, 2010.
31. Nourbakhsh, S. A.; Jaumotte, B. A.; Hirsch, C.; Parizi, H. B. *Turbopumps and Pumping Systems*; Springer, 2008.
32. Rossmann, L. A.; Grayman, W. M. Scale-Model Studies of Mixing in Drinking Water Storage Tanks. *J. Environ. Eng.* **2000,** *125* (8), 755–761.
33. Farmani, R.; Savic, D. A.; Walters, G. A. *The Simultaneous Multi-Objective Optimization of Anytown Pipe Rehabilitation, Tank Sizing, Tank Siting and Pump Operation Schedules, Critical Transitions in Water and Environmental Resources Management*; ASCE, American Society of Civil Engineers: Salt Lake City, 2004; pp 4663–4672.
34. Todini, E. Looped Water Distribution Networks Design Using a Resilience Index Based Heuristic Approach. *Urban Water* 2000, *2* (3), 115–122.
35. Prasad, T. D. Design of Pumped Water Distribution Networks with Storage. *J. Water Resour. Plann. Manag. ASCE* **2010,** *136* (1), 129–132.
36. Munshi, J. A. *Rectangular Concrete Tanks*; Portland Cement Association, 1998.
37. Munshi, J. A. *Circular Concrete Tanks Without Prestressing*; Portland Cement Association, 1993.
38. Munshi, J. A. *Design of Liquid Containing Concrete Structures for Earthquake Forces*; Portland Cement Association, 2002.
39. NCHRP Report 611. *Seismic Analysis and Design of Retaining Walls, Buried Structures, Slopes, and Embankments*; American Association of State Highway, National Cooperative Highway Research Program, 2008.

40. Nourbakhsh, S. A.; Jaumotte, B. A.; Hirsch, C.; Parizi, H. B. *Turbopumps and Pumping Systems*; Springer, 2008.
41. NSF/ANSI 61. Drinking Water System Components—Health Effects; NSF National Standard/ANSI National Standard, 2010.
42. NZS 3106. *Design of Concrete Structures for the Storage of Liquids*; Cement and Concrete Association of New Zealand, Standards Association of New Zealand, 2009.
43. Asli Kian, H.; Asli Kaveh, H. Isolated Pressure Zones Based on GIS as a Solution for Water Network Problems. *J. Water Practice Technol.* 2023, *17* (9), in press.
44. https://iwaponline.com/wpt/article/doi/10.2166/wpt.2022.119/91338/Isolated-pressure-zones-based-on-GIS-as-a-solution
45. Yeung, H. Modelling of Service Reservoirs. *J. Hydroinf.* **2001,** *3* (3), 165–172.
46. Asli, H. H. *Reclaimed Asphalt Pavement (RAP) Based on Geospatial Information System (GIS) and Networked Sensors Modelling, New Advances in Material Technologies Experimental Characterizations, Theoretical Modeling, and Field Practices*, in Press by the Apple Academic Press, Inc., 2023. https://www.appleacademicpress.com/new-advances-in-material-technologies-experimental-characterizations-theoretical-modeling-and-field-practices/9781774914847
47. Asli, H. H. Applications of Networked Sensors and the Internet of Things (IoT) for Wastewater Treatment, Sustainable Water Treatment and Ecosystem Protection Strategies, in Press by the Apple Academic Press, Inc., 2023. https://www.appleacademicpress.com/sustainable-water-treatment-and-ecosystem-protection-strategies-/9781774915189
48. Asli, H. H. *Modeling of Corrosion for Water System by Networked Sensors and the Internet of Things (IoT) in Compliance with Geography Information System (GIS), Sustainable Water Treatment and Ecosystem Protection Strategies*; in Press by the Apple Academic Press, Inc., 2023. https://www.appleacademicpress.com/sustainable-water-treatment-and-ecosystem-protection-strategies-/9781774915189
49. Gualtieri, C. Analysis of Flow and Concentration Fields in a Baffled Circular Storage Tank. *XXXIII IAHR Congress*, Vancouver, Canada, August, 10/14, 2009; pp 4384–4392.
50. MacPherson, G.; Cripps, S.; de Schaetzen, W. Storage and Pumping Capacity Analysis Using Hydraulic Modeling. In: *Conference Proceeding Paper, A. A. Balkema Publishers, Water Management Challenges in Global Changes*; CCWI: London, 2007; pp 439–446.
51. Teixeira, E. C.; Siqueira, R. N. Performance Assessment of Hydraulic Efficiency Indexes. *J. Environ. Eng.* **2008,** *134* (10), 851–859.
52. Sturman, R.; Wiggins, S. Eulerian Indicators for Predicting and Optimizing Mixing Quality. *New J. Phys. 11* (6), 75031.
53. Patwardhan, A. W. CFD Modeling of Jet Mixed Tanks. *Chem. Eng. Sci.* **2002,** *57*, 1307–1318.
54. Martínez-Solano, F. J.; López-Jiménez. P.A; Iglesias-Rey, P.; Pérez, R. 2006. Utilización de la fluidodinámica computacional (CFD) como herramienta para la determinación de parámetros de los depósitos en modelos de calidad del agua" (in spanish). V Congreso Ibérico sobre gestión y planificación del agua. Universidad de Algarbe, Faro, Portugal.

CHAPTER 5

Polyhydroxyalkanoate Production in Wastewater Treatment Plants: Process Scheme, Operating Conditions, and Potential Analysis for Sustainable Water Resources Management

NEGAR SARAEI

Department of Civil and Environmental Engineering, Lamar University, Beaumont, TX, USA

ABSTRACT

In the production of polyhydroxyalkanoates (PHAs) by using pure culture, many studies have been done so far. This work investigated the production of PHA through limited oxygenation in the activated sludge cultivation process in the anaerobic stage. In this work, the limited injection of pure oxygen in the anaerobic stage increased the production of PHAs from activated sludge. There were effluents that had a significant percentage of volatile fatty acids (VFAs). Wastewater from cellulose acetate factories (with an average of 1200 mg L^{-1} of acetic acid) was a suitable composition for the production of PHAs. A significant amount of VFAs was produced by fermenting different wastes, including urban sewage. The result of this work showed that urban wastewater can be used as a cheap source for the production of PHAs. It can also be used for the production of bioplastics. These types of plastics have properties that are comparable to those of synthetic plastics.

Sustainable Water Engineering: Smart and Emerging Technologies. Ali Pourhashemi, Kian Hariri Asli, Ann Rose Abraham, & A. K. Haghi (Eds.)

5.1 INTRODUCTION

With the increase in population and in line with the increase in consumption in the communities, as well as the increase in the level and lifestyle, the daily consumption of all kinds of plastics produced by petrochemical units has increased. Plastics consumption includes the use of all kinds of plastics in food packaging and also the use of plastic bags for moving and carrying garbage. Since these types of plastics are nondegradable in nature, they remain unchanged in the environment for many years. Plastics consumption is considered one of the main problems in solid waste disposal centers. It should be mentioned that every year a considerable amount of money is spent by the relevant organizations for the disposal of these types of plastics that enter urban and industrial wastes in different ways, and it is considered one of the problems of solid waste disposal management.

Due to the increasing consumption of these types of substances and the warnings of various environmental organizations regarding the problems caused by their release in nature, for the past two decades, worldwide research has been started in order to make similar compounds that are also biodegradable. So far, a large number of degradable plastics have been produced around the world, which is also compatible with the environment. But today, their mass production is limited due to factors such as the relatively high price of the substrate, the low production rate, and also the costs related to keeping the environment sterile (to prevent the contamination of microbes in pure culture). The group of polymers cannot compete with other plastics. According to one of the studies conducted, the significant cost of raw materials is the main reason for the high price of degradable plastics compared to normal plastics. In a 1996 report, Lee estimated the cost of producing degradable polymers in pure culture to be more than the cost of synthetic plastics.

Therefore, finding a compound with the physical characteristics of plastics that can be decomposed in nature and is also economically competitive with ordinary plastics is considered a suitable alternative.

Among the degradable plastics that have been prepared so far, a group called polyhydroxyalkanoates (PHAs) and its copolymers, which have more than 40 types of compounds, are notable due to their complete degradability of 100%, low production cost, and also the relative simplicity of the process compared to other polymers. Hence, biodegradable plastics have been given more attention. The PHAs are polyester hydroxyalkanoates, and their molecular weight is in the range of 2×10^5 to 3×10^6 Da, depending on the type of microorganism and growth conditions. These types of compounds are used even in surgical operations due to their physical and mechanical properties and also due to their compatibility with the body.[91]

As mentioned, the most important feature of PHA is its complete biodegradability. This composition can be decomposed by aerobic, nonaerobic, and amphibian microorganisms, and it turns into CO_2 and water vapor.

A group of bacteria and also a group of filamentous bacteria have the ability to make these types of polymers or their copolymers as a source of carbon storage inside the cell.[92] PHA is accumulated in the form of granules inside the cells of microorganisms and can be easily identified by staining with black Sudan or blue indigo azure.

Recombinant rabble to synthetic plastics has 100% water resistance, a low production cost compared to other degradable plastics, and flexibility.

During the last two decades, many efforts have been made to produce optimal PHAs, but despite these efforts, the current price of these types of compounds is still several times the cost of ordinary plastics.

In general, the factors affecting the price of PHA production can be considered in the form of four parts, including yield, production efficiency, carbon source price, and finally the costs related to extraction and purification.

PHA production using mixed cultivation, such as activated sludge, and also the use of low-cost carbon sources, followed by process optimization to increase yield and production efficiency, are among the measures that can reduce PHA production costs.

In relation to the production of PHA using pure culture, many studies have been conducted, and the necessary conditions for the optimal production of polymers have been determined. But regarding the production of PHA by mixed cultivation activated sludge, due to the novelty of the subject, the studies have not yet been completed, and more comprehensive measures are needed.

In 1998, Sato et al.[108] described the production of polymers using activated sludge when sodium acetate was used as a feed carbon source. In this research, by using the microaerophilic-aerobic system through limited oxygenation in the anaerobic phase, the increase in the production of PHAs from activated sludge has been experienced. It should be mentioned that in industries, there have been effluents that have a significant percentage of volatile fatty acids (VFAs), among which the wastewater from cellulose acetate production factories has an average of 1200 mg L^{-1} of acetic acid. It is mentioned that they are a suitable combination for the production of PHAs too. A significant amount of VFAs can be produced by fermenting different wastes, including urban sewage, and can be used as a cheap substrate in the production of PHA.

With regard to the above description and taking into account the problems related to the disposal of plastics used in product packaging as well as waste transportation and considering the costs related to the disposal of excess sludge from sewage treatment plants, the production of PHAs using activated sludge and the determination of factors effective have been chosen as the subject of this research.

For the first time in 1925, a microbiologist named Lemoigne discovered and described polybetahydroxybutyrate (PHB) at the Pasteur Institute in Paris. He observed granular bodies in the cytoplasm of *Bacillus megaterium* that were not soluble in ether. He realized that these compounds have the general formula $(C_4H_6O_2)n$ and are polyesters. Due to the difference in the melting point of this compound, Lemoigne realized that this polymer has two parts. He separated the polymer with the help of chloroform.[41]

About 30 years later, PHB was again noticed in 1952. Capps and Peudelnoel[12] observed that both the polyester components isolated by Lemoigne hydrolyzed products were high-molecular-weight linear polyester whose melting point is 180°C.

In 1958, Williamson and Wilkinson[12] reported data on the molecular weight and physical properties of PHB. In the same year, Makraei and Wilkinson[41] observed that intracellular PHB accumulation increased with nitrogen limitation in culture medium.

In the late 1950s and early 1960s, Werber and Baptiste[12] in a company called WR Grace and another company in the United States of America began to produce small amounts of PHB for commercial purposes. They obtained the franchise to increase the production and isolation of PHB. Of course, their fermentation method had low efficiency, and the method of extracting PHB from inside the cell using solvent was expensive for them. In addition, the produced polymer was contaminated with the remains of bacterial bodies during separation, which made the melting process difficult. The production project of the aforementioned company was abandoned due to the mentioned problems, and the industrial production of this polymer was postponed for a decade. It is worth mentioning that this project became a precursor for further research on PHB as a biodegradable plastic compatible with vital systems.

In 1968, ICI in England began to develop single-cell protein (SCP) technology. SCP food products named protein were successfully produced. Of course, the initial price of the substrate for bacterial fermentation was much higher than the limit to be able to provide the protein at a commercial level. In this way, this project was abandoned and left the way to study on PHB. On the other hand, the technical information about fermentation on a large scale and the skill in the field of polymer processes increased significantly. In this opportunity, ICI provided the necessary equipment for the commercial production of PHB. ICI found the bacterium *Alcaligenes eutrophus*, which was able to accumulate PHB to the extent of more than 70% of its dry weight. Pure PHB did not have any special advantage in comparison with polypropylene due to its brittleness. But PHB copolymers had more interesting properties and were therefore more suitable for industrial applications.[41]

Therefore, Holmes and coworkers in 1981 at ICI presented a controlled fermentation process that caused the bacterial production of P(3HA) copolymers from different carbon sources. This copolymer, under the name "Biopol," was produced from *Alcaligenes eutrophus* grown on propionic acid and glucose. A change in the ratio of propionic acid and glucose caused a change in the ratio of copolymer components, which finally led to the creation of copolymers with different melting points and mechanical and thermoplastic properties.[41]

Finally, in 1990, the Wella company in Germany produced the first Biopol commercial product in the form of a biodegradable bottle for shampoo packaging.[41]

In another research in 1997, the production of bioplastics from the surplus sludge obtained from urban sewage treatment was studied.[72] In this research, which was carried out on a laboratory scale (PHAs), the sludge obtained from urban sewage treatment was recycled in two stages. In the first stage of mud in the face anaerobic digestion in thermophilic conditions and then PHAs were produced from organic compounds in the supernatant by the microorganism *Alcaligenes eutrophus*.

In this work, the sludge used has 3–5% dry solids and 50–70% volatile solids, and it has been fermented into fatty acids and other compounds at a temperature of 50–65°C in anaerobic conditions. In this research, the reduction of volatile solids and the production of soluble organic compounds at different temperatures, pHs, and retention times have been investigated. In this research work, it has been concluded that more dissolved organic substances are formed at a temperature of 65°C and a pH > 5.5. In this study, the amount of PHAs produced was about 34% of the weight of the cell mass, which showed a 1% increase compared to when VFAs were used in pure form as a carbon source.

In this process, about 78% of the organic carbon in the supernatant has been used. According to the investigations, the conversion of four main acids in the saliva, including acetic acid, propionic acid, butyric acid, and valeric acid, was 87.6%, 62.6%, 56.8%, and 32%, respectively. The PHAs produced in this method contain (74% by weight) C4 monomers, and their melting point is 167°C, that is, 9°C lower than PHB poly-3-hydroxybutyrate. The bioplastic produced in this research was decomposed in the soil environment during 5 weeks to the extent of 22–27%, and when it was floated in the relevant sludge during 6 weeks in anaerobic conditions, 70% of the cases analyzed were aerobic.

In 1997, the production of PHA from palm oil was studied using the microorganism *Rhodobacter sphaeroides* (IFO 12203) from every 15 g of organic acid. About 2 g of PHA was produced; more than 60% of the dry weight of PHA sludge has been produced.[46]

In 1997, Chanak et al.[25] studied the effect of hydraulic retention time and also dissolved oxygen in order to remove nutrients in the AS-Biofilm combination process and concluded that phosphorus removal in anaerobic-anoxic and aerobic stages is proportional to the amount of PHA in the product. This researcher also conducted experiments in 1998 to compare phosphorus removal and PHA production in one system (anaerobic-anoxic-aerobic) on a pilot scale and obtained similar results.

In 1998, the effect of carbon sources on the production of PHAs during the biological removal of phosphorus was studied by Lemo et al.,[74] whose main goal was to optimize phosphorus removal and polymer production. In this study, acetate, propionate, and butyrate were used as carbon sources, and it was concluded that the composition of the produced polymer varies depending on the type of substrate. For example, acetate tends to produce a copolymer that is a combination of hydroxybutyrate (HB) and hydroxyvalerate (HV), and usually HB is dominant in it. However, propionate tends to produce more HV and the amount of HB in it is negligible. Finally, when butyrate is used, the ratio of HB to HV is reported to be around 1.5. In this research, the efficiency of polymer production compared to the carbon used in cases where the carbon source was acetate, propionate, and butyrate was reported as 0.97, 0.61, and 0.21, respectively.

In 1999, Chu and Yu investigated the production of degradable plastics from chemical wastes.[21] In this research, *Alcaligenes* is named as a microorganism that is able to store PHAs inside the cell. In this work, the amount of polymer production increased with the increase of C to N, and the highest amount of polymer was produced in C to N, about 100. In this research, while producing bioplastics, the volume of excess production sludge that needed to be stabilized has also been reduced. Polymer stored by the microorganism *Alcaligenes* spp. mainly includes poly-beta-hydroxy butyric acid and copolymers of beta-hydroxybutyric acid 3-HB and beta-hydroxyvaleric acid 3-HV. The wastewater used in this research has been selected for xenobiotic-branched carboxylic acids and ketones.

In 1999, the production of PHAs from food industry wastewater was studied by Yu et al.[24] In this research, the wastewater from the beer factories and soybean processing factories was used, and it was determined that there is a possibility of producing PHAs in the activated sludge processing food industries as a source of carbon. It is possible to prepare polymers with the required physical properties.

In 1999, in a research work, the preparation of PHAs from wastes was introduced as a process to reduce the amount of organic wastes.[71]

In 2000, the production of PHAs using activated sludge was studied by Takabatek et al.[124] In this research, the control of production polymer compositions as well as the enrichment of biomass from polymers has been considered. In this study, it was concluded that the quality of the produced product can be controlled by controlling the composition of wastewater.

In 2001, the kinetics of the combined consumption of VFAs in the process of PHA production by the bacterium *Ralstonia eutropha* was investigated.[58] In this research, the rate of PHA production from acetate, propionate, and butyrate, individually or in combination, was investigated. In the best conditions, with a combination of acetic acid and butyric acid, the rate of PHA production was about 60 mg per gram of biomass production per hour. In this research, when the combination of acetic acid and propionic acid was used, the rate of PHA production was reduced to 35 mg per gram of biomass produced per hour.

In 2003, Adeline et al.[1] studied the effect of acetate concentration on pH and sludge cell retention time during the production of PHA from activated sludge. In this research, pH between 7 and 8 has been determined as the most appropriate level for maximum polymer production in the sludge application stage. In this experiment, increasing SRT caused the ability of sludge to produce polymers to be reduced. According to the results of this research, the addition of acetate to the wastewater during the sludge application stage has increased the amount of polymer produced by 10%.

5.1.1 POLYMER STORAGE IN MICROORGANISMS

Based on the research done,[29] four groups of polymers, including lipids, carbohydrates, polyphosphates, and nitrogen compounds, were stored in microorganisms as nitrogen.

Some microorganisms are able to store more than one type of polymer in their cells; among them, the polyphosphate-accumulating organisms (PAOs) are used in the BPR system. These types of microorganisms are able to store glycogen, PHA, and polyphosphate.

Dawes and Senior[28] stated that the environmental conditions and the mechanisms used will affect the composition of the stored polymer. In addition, the carbon source used will also affect the type of polymer stored in the cell. For example, when the source of carbon supply is changed from glucose to acetate, the amount of carbohydrates in *E. coli* is reduced and the amount of stored lipids is increased.

According to the studies of Sasikala and Ramana,[106] when the carbon source is metabolized through acetyl-CoA and no pyruvate is produced as an intermediate, usually the carbon is directed toward the production of PHA. But if this carbon source is metabolized through pyruvate, the amount of glycogen in the cell will be higher than other compounds.

Dawes[29] stated that the important role of polymers stored by microorganisms is to supply the carbon and energy required by the microorganism. It allows microorganisms to use it as a source of carbon and energy when there is a nutrient limitation. Therefore, microorganisms that do not have the ability to store polymers, when faced with the limitations of food, tend to use cellular content such as protein and RNA.

5.1.2 LIPIDS PHA (POLYHYDROXYALKANOATES)

In this section, there are some materials related to the production of PHA by microorganisms and the factors affecting their production and composition.

Researchers[4,15,35,69,71,106,120] have done extensive studies on the production of PHA by microorganisms. This type of compound can accumulate inside the cells of microorganisms.

They can be produced by a wide variety of bacteria. Among the most important cases, it can be mentioned the *Ralstonia eutropha*. It was previously known as *Alcaligenes eutrophus*. Due to the high ability of this type of bacteria to store PHA, this type of microorganism has been the subject of much research published so far. It should be mentioned that this type of bacteria is able to store PHA up to about 80% of the dry weight of the cell.[71] There is a list of a number of microorganisms that are able to store PHA inside their cells.

According to the research of Lafferty et al.,[69] PHB has properties such as tensile strength and flexibility similar to those of polystyrene and polyethylene. Polymers produced from PHA are completely 100%

degradable in both aerobic and anaerobic conditions,[89] which is one of the advantages of this group of polymers in comparison with degradable plastics that are produced artificially.

The PHAs, in terms of construction composition, can have different types. Although more than 80 different types of PHAs have been identified inside the cells of microorganisms,[71] usually the PHAs extracted from the cells of microorganisms are mainly composed of PHB and polybetahydroxyvalerate (PHV).

Today, in the world, two types of PHAs, namely homopolymer PHB and copolymer 3HB-3HV, are produced for industrial use.

In a study, Lafferty et al.[69] mentioned asymmetric growth conditions such as nutrient limitations (nitrogen, sulfate, and phosphorus compounds), low oxygen concentration, or a high carbon-to-nitrogen ratio in the substrate as accelerating factors in polymer production. In another study by Sasikala and Ramana,[106] in addition to limitations caused by nitrogen, phosphorus, sulfate, and oxygen, limitations in iron, magnesium, manganese, potassium, and sodium were also found effective on the PHA production process.

According to the second research,[35] when the growth conditions are unbalanced, acetyl-CoA cannot enter the tricarboxylic acid (TCA) cycle to obtain energy for the cells. Due to the high concentrations of NADH, it prevents the synthesis of enzymatic citrate, one of the key enzymes of the TCA cycle, and as a result, increases the amount of acetyl-CoA. Acetyl-CoA is used as a substrate for the production of PHA during three consecutive enzymatic reactions.

Also, high concentrations of intracellular CoA prevent the action of 3-ketothiolase enzyme, one of the three effective enzymes of PHA biosynthesis.

If there is no obstacle to the entry of acetyl-CoA into the TCA cycle, free CoA is released as an acetyl moiety from the activity of citrate synthesis. For example, when acetyl-CoA is consumed, the concentration of intracellular CoAs increases, and as a result, PHA production is prevented.

PHA can act as a source of energy or carbon for microorganisms during periods of food limitation.

In enhanced biological phosphorus removal (EBPR) systems, PHA is produced and stored by PAOs in the anaerobic phase in conditions where there is no electron acceptor for growth mechanisms. Such a process allows EBPR systems to sequester existing substrates as well as microorganisms that do not have the ability to store phosphorus and PHA. According to Lee's studies,[71] so far, four different biological synthesis routes have been determined for the production and storage of PHAs, which are mentioned below.

5.1.3 PATHS OF PHA BIOSYNTHESIS

As mentioned, four biological synthesis routes have been determined for the production of PHA.

5.1.3.1 PHA BIOSYNTHESIS ROUTE USING R. EUTROPHA

Most microorganisms are able to produce PHA by using this route. Among them, the production of PHA by *R. eutropha*, *Zoogloen ramigera*, and *Azotobacter beijerinckii* can be mentioned.[35] In this route, the substrate is first converted into acetyl-CoA. Then 2 moles of acetyl-CoA are used, and it turns into 1 mole of PHB. In this way, acetyl-CoA is subjected to three consecutive enzymatic reactions to produce PHB.[4,35]

In this process, if propionic acid is used as a substrate alone, a PHB-PHV copolymer will be obtained. In this case, acetyl-CoA is formed by removing the carbonyl carbon from propionyl-CoA. Then, 2 moles of acetyl-CoA are used to produce one HB unit of copolymer, while one HV unit is formed from the reaction of acetyl-CoA and propionyl-CoA.[35]

According to studies,[35] simultaneously with the synthesis of PHA by *R. eutropha*, if there is a limitation in nitrogen, there is also a possibility of PHA degradation, which is called the natural cycle of PHA metabolism.

For example, when *R. eutropha* with about 56% PHA is exposed to butyric acid as the only substrate under nitrogen limitation conditions, the composition of PHA inside the cell changes significantly; so after about 48 h, the amount of PHV decreases from 56 to 19%. These findings indicate the simultaneous synthesis and consumption of PHA, which is the natural cycle of PHA metabolism.

5.1.3.2 THE PATH OF PHA BIOSYNTHESIS USING RHODOSPIRILLUM RUBRUM

This path is similar to the path of PHA biosynthesis using *R. eutropha*, with the difference that in this path, two units of enoyl-CoA are also converted into L-3-hydroxybutyryl-CoA. In the second step, D-3-hydroxybutyryl-CoA will be involved in the vicinity of crotonyl-CoA.[4,35,71] The simple diagram of this biological pathway can be considered as follows (eq 5.1):

$$\begin{aligned}&\text{acetate} \rightarrow \text{acetyl CoA} \rightarrow \text{acetoacetyl CoA} \rightarrow \text{L}-3\\&\quad -\text{hydroxybutyryl}-\text{CoA} \rightarrow \text{Crotonyl CoA} \rightarrow \text{D}-3\\&\quad -\text{hydroxybutyryl}-\text{CoA} \rightarrow \text{PHB}\end{aligned} \tag{5.1}$$

5.1.3.3 PHA BIOSYNTHESIS PATHWAY USING PSEUDOMONAS OLEOVORANS

This biosynthetic pathway is true for *P. oleovorans* and most pseudomonads from rRNA homology group I.[71] These organisms produce medium-chain PHA (C6-C9) from medium-chain length alkanes, alcohols, or alkanoates. According to the research,[35] the production of PHA with short-chain length, such as PHB-PHV copolymer and PHB homopolymer, can also be the result of the action of this group of microorganisms. But in this case, the amount of production decreases to less than 1.5%.

5.1.3.4 PHA BIOSYNTHESIS ROUTE USING P. AERUGINOSA

Most pseudomonads from rRNA homology group I, except *P. oleovorans*, can also produce medium-chain PHA using this method. The route used by these organisms is called *P. aeruginosa* route. Also, Steenbuchel[120] concluded that medium-chain PHAs that are saturated by this route are obtained from unrelated substrates such as gluconate or acetate.

5.1.4 FACTORS AFFECTING THE PRODUCTION OF PHA AND ITS COMPOUNDS

As mentioned before, the type of substrate used will affect the percentage of PHB and PHV in the polymer. Based on the studies,[35] the polymer compositions produced by *R. eutropha* will be PHB homopolymers when alkanoic acids with even carbon numbers are used as substrates. If the number of carbons in the alkanoic acid chain is odd, a PHB-PHV copolymer will be produced.

In this case, when fructose or glucose is used as a substrate, PHB homopolymer is produced.

Based on the research done, the concentration of the used substrate has also been reported to be effective in terms of the amount of polymer produced. For example, when propionate was used as a substrate for the production of PHA from *R. eutropha*, reducing the concentration of propionate from 14 to 2 g/L caused a decrease in the amount of PHA produced from 56 to 12%.

Doi et al.[34] studied the composition of the copolymer produced using *R. eutropha* H16 when a combination of acetate and sodium propionate was used as a substrate. In this research, increasing the concentration of sodium propionate showed an increase in the percentage of PHV in the copolymer. Also, when sodium propionate was used alone as a substrate, copolymer 3HB-3HV was produced.

In 1981, the British ICI Institute investigated the production of 3HB-3HV copolymer from *R. eutropha* when propionic acid and glucose were used as carbon sources. In this research, the molar percentage of PHV in the produced copolymer depends on the composition of the substrate used in such a way that when pentanoic acid and butyric acid are used as carbon sources, the molar percentage of PHV is more than 95% increased. Therefore, the composition of the produced copolymer, as well as its physical properties and thermal behavior, can be controlled by changing the substrate used.[36] They further conducted experiments using different carbon sources to produce PHA using *R. eutropha*. In these experiments, the use of different carbon sources has resulted in the production of different PHA compositions.[36]

In this study, the biodegradation of a thin film of produced PHA (with a weight of 4–8 mg, dimensions of 10 × 10 mm, and a thickness of 0.03–0.06 mm) has been investigated. Based on the tests, the rate of decomposition of the copolymer 3HB-4HB is higher than other tested samples. In this experiment, 3HV is also mentioned as a factor reducing the rate of copolymer decomposition.

Shimiz et al.[115] studied the production of PHA from *R. eutropha* H16 using valeric and butyric acid. They reported the optimal conditions for PHB production using this bacterium at a concentration of 3 g/L of butyric acid and a pH equal to 8. In this study, no information was provided regarding PHV or other PHAs. In the above conditions, the amount of PHB stored inside the microbe was reported as 75%. In this experiment, by changing the concentration of butyric acid, the amount of PHB produced decreased. In this experiment, the use of concentrations of 0.03, 0.3, and 10 g/L of butyric acid produced 44%, 55%, and 63% of PHA, respectively. Following this, the experiment was repeated with a concentration of 3 g/L of butyric acid at pHs of 7.5 and 6.9, in which the amount of PHA production was reported as 58% and 53%, respectively. In this experiment, increasing the pH above 8.4 stopped the production of PHB. In this research, when propionic and butyric acids were used as carbon sources, the 3HB-3HV copolymer was obtained. But the amount of polymer production was reduced compared to when 3HB was produced using butyric acid alone.

In another research,[119] the production of PHB-PHV copolymer using *R. eutropha* strain R3 under nitrogen limitation conditions was studied. In this research, 47%, 35.7%, 29.5%, and 43.2% of polymer (PHA) were produced by using fructose, gluconate, acetate, succinate, and lactate, respectively. In this test, the amount of PHV in the produced copolymer was reported in the range of 4–7%. Also, when the above test was done with magnesium or sulfur limitation and fructose was used as a carbon source, the amount of PHA produced was 45% and 47%, and the amount of PHV in the copolymer was reported as 7% and 6%, respectively. In the following, when gluconate is used as a substrate, the amount of PHA production in magnesium limitation conditions is 33% and in sulfur limitation conditions, it is about 38%, and the amount of PHV in each of the mentioned conditions is 4% and 5%, respectively.

In order to use a cheaper carbon source, Burke et al.[10] investigated the production of PHA by 118 methylotrophic microorganisms on methanol. In this research, a significant amount of PHA was produced by using a mixture of methanol and valerate and the microorganism *Methylobacterium extorquens*. In this research, the amount of PHA produced was in the range of 60–70%, and the amount of PHV in the copolymer was reported to be around 20%.

In another study, Burke et al.[11] investigated the production of PHA with the help of *Methylobacterium extorquens* ATCC 55366 and methanol as a source of carbon and energy in a discontinuous fermentation system. In this system, the amount of PHB production has been reported in the range of 40–46% based on the dry weight of the sludge. In this system, the production of biomass and the growth rate of microorganisms are affected by the mineral compounds in the system in such a way that there is an absence of ammonium sulfate or manganese sulfate and an absence of a combination of calcium chloride, bivalent iron sulfate, magnesium sulfate, and zinc sulfate. It has reduced the production of biological mass. In this research, high concentrations of ammonium sulfate were introduced as toxicity to the above system, while increasing the concentration of $MgSO_4$ and $FeSO_4$ and the mixture of micronutrients (trace elements) increased the growth rate of microbes. In other words, the maximum cell density of *M. extorquens* has been achieved when a suitable ratio of mineral compounds has been used.

In a study by Suzuki et al.,[122] the highest PHB production of dry cell weight was obtained using 66% *Pseudomonas* SPK when the only source of carbon used was methanol. In this study, in order to achieve high concentrations of PHB, the substrate has been tested in different combinations. It should be mentioned that phosphate and ammonium were used in low concentrations in this experiment. In this research, nitrogen deficiency has been reported as the most effective factor to accelerate PHB storage.

Also, the limitation in dissolved oxygen concentration has caused a decrease in the growth rate of the biological mass, and as a result, a decrease in the production of PHB, which is in contradiction with the results reported by others.

In another research,[69] using *R. eutropha* and *Azotobacter beijerinckii*, PHB was produced under conditions of oxygen limitation, and it was concluded that oxygen limitation accelerates the rate of PHB production.

In this research, the same PHB with a melting temperature of 176°C and a molecular weight of 105°C was obtained from *Pseudomonas*.

Daniel et al.,[27] in a culture medium where feed methanol was added intermittently using *Pseudomonas* 135 under conditions of ammonium limitation, experienced the production of 55% polymer (PHB). Further, when the microbes were placed in environments with a limit of Mg^{2+} and PO_4^{3-}, the amount of PHB production decreased to 42.5% and 34.5%, respectively. PHB produced by this group of microorganisms has a melting point of about 173°C.

Bayrom[18] from the ICI Institute has stated his experiences in relation to the industrial production of PHA as follows:

- Since *Ralstonia* spp. is easily able to produce high-molecular-weight PHA, it is therefore the selected microorganism of the ICI Institute.
- The production of PHA from *Azotobacter* and methylotrophs has also been studied. In the case of methylotrophs, the polymer produced has low production efficiency and molecular weight. It is difficult to separate the polymer from methylotrophs. *Azotobacter* was also not interesting in this regard because this organism used the carbon source instead of producing more PHA for the production of polysaccharides.
- *R. eutropha* is able to produce 70–80% of polymer under conditions of phosphate ion limitation. In this case, when a mixture of glucose and propionate was used as a carbon source, the microorganisms were able to produce copolymers.
- In another study,[11] the main problem in the use of different types of *R. eutropha* was reported to be the inefficient use of propionate. For example, only about one-third of this material enters the HV part of the production copolymer. In another research, the mutant strain Bs-1 was reported to be more effective in consuming propionate, and in this case, more than 80% of propionate was reported in the HV part of the production copolymer.

According to the studies,[54] *Alcaligenes latus* is able to store up to 80% of PHA in its cells under normal growth conditions. Therefore, the one-step process of PHA production can be used with this organism. According to Doi's research,[35] a two-stage process with noncontinuous feather feeding is the most commonly used technique to produce maximum PHB and high cell concentrations. In this method, in the first stage, which is also called the growth phase, the maximum amount of biological mass (biomass) is produced by using the appropriate composition of the substrate. Then, by limiting one of the nutrients, PHA production is accelerated in the second stage, or the polymer storage phase.

In 1996, Yemen et al.[129] investigated the production of PHA by *A. latus* using sucrose as a carbon source. In this case, the high cell concentration (142 g/L) was obtained in 18 h, and the PHB concentration was recorded at the end of the mentioned time at a rate of 50%. Therefore, they concluded that the size of the inoculum reduces the cultivation time. They also measured the time required to perform the PHB production process using *R. eutropha* with glucose as a carbon source in the same conditions as the previous experiment. In this experiment, the time needed to reach the cell concentration of 122 g/L was about 30 h, and during this period, the concentration of PHB inside the cell increased to about 65%. The average molecular weight of PHB produced in this research was 3.6.

In 1996, Renner et al.[100] studied the production of copolymers using 13 different bacteria from the third group of the rRNA family (rRNA superfamily III). They concluded that different bacteria in the same conditions will be able to produce PHAs with different ratios of PHB-PHV.

In 1990, Anderson et al.[5] studied the production of copolymers using Nocardia, Rhodococcus, and Corynebacterium groups. They concluded that these species have differences in polymer storage and composition under the same conditions.

Therefore, in summary, it can be said that the amount and composition of PHAs are mainly influenced by the type of microorganisms, the type of carbon source substrate used, the concentration of the substrate, and the environmental growth conditions that affect PHA production by different microorganisms under different growth conditions.

5.1.5 PHYSICAL PROPERTIES OF POLYHYDROXYALKANOATES (PHAS)

As mentioned earlier, different types of microorganisms produce polymers, mainly known as poly-b-hydroxyalkanoates (PHAs), as carbon and energy reserves inside their cells. The reason for this naming is the presence of different alkyl groups in position b. The PHB is also one of the polymers related to this group, which has properties similar to polypropylene (PP). The three main characteristics of PHB, that is, thermoplastic performance, 100%

resistance to water, and 100% biodegradability, have made this composition more attractive to environmentalists today.

In 1994, Bouma et al.[9] named PHB as an aliphatic homopolymer with a melting point of 179°C, which has high transparency (80%) and decomposes at a temperature higher than its melting point.

According to the research,[31] if PHB is kept at 190°C for 1 h, its molecular weight will be reduced to half of its initial value.

PHAs can show different physical properties depending on their compositions. The range of these physical properties varies from brittle and unstable to heat to soft, firm, and durable.[135]

One of the effective factors in this relationship is the PHV/PHB ratio. The physical properties of PHB, such as crystallization ability and tensile expansion ability, depend on its molecular weight, which is influenced by the strength of the microorganism used, the growth conditions, and the degree of purity of the sample. In 1990, Byrom[17] summarized the two main advantages of the PHB-PHV copolymer over PHB homopolymer as follows:

First, the copolymer has a lower melting point. Although the HV present in the copolymer composition obviously does not reduce the temperature to the extent where molecular weight reduction occurs, this important feature allows the copolymer to be prepared in a larger range of temperature conditions.

Secondly, the PHB-PHV copolymer has a lower degree of crystallization, which increases its flexibility and durability. One of the disadvantages of the low crystallization power is the lengthening of the process time.[16]

In 1994, Bouma et al.[9] suggested that the composition of PHV can be changed to obtain a suitable composition. Based on the results of this research, materials with a high PHV composition are soft, firm, and durable. While compounds with lower PHV content are hard and brittle (in the product produced by the ICI company under the name Biopol, the amount of PHV in the PHB-PHV copolymer ranges from 0 to 30%).

In the studies conducted by Lee in 1996,[71] the molecular weight of PHA was reported in the range of 10^5–10^6. In this research, it is also noted that PHB with a higher molecular weight is more suitable for industrial use.

In 1995, Burke et al.[11] found that the molecular weight of PHB can be reduced in the polymer production stage. In addition, in the biomass extraction process, the molecular weight of PHB may decrease.[67]

The research also shows that when the concentration of butyric acid increases, the molecular weight of PHB decreases,[115] but pH has an insignificant effect on the average molecular weight of PHB. In this research, the highest molecular weight of PHB, equal to 3.3×10^6, was obtained at a concentration of 0.13 g/L of butyric acid. While the molecular weight of 2×10^6 was obtained in the optimal conditions of PHA production (concentration of 13 g/L butyric acid at pH = 8). In this study, the molecular weight was reported in the range of 1.6×10^6–3.3×10^6.

5.1.6 APPLICATIONS OF PHAs

The research conducted on PHAs has shown the biodegradability, compatibility with vital systems, polyester nature, and thermoplastic properties of this polymer. In addition, PHAs are produced from renewable sources, which justify the many uses of PHAs.[12]

HB and HV units have photoactive carbon, and both have D (−) spatial form. This property of polymers has been used for the chemical synthesis of photoactive substances that have biological activity only in a specific spatial form, such as some drugs. This polymer can also be used in chromatography to separate optical isomers.

In general, the applications of PHAs are concentrated in three main areas: medicine and pharmaceuticals, agriculture, and product packaging.

5.1.6.1 USE OF PHAS IN AGRICULTURE

In agriculture, to protect wheat from soil pests, you can use a suitable insecticide that is covered with PHB. In this case, granules of PHB containing insecticide are also planted along with sowing in the autumn. The activity of PHB-destroying bacteria causes the gradual release of the insecticide, and thus the seeds germinate in an environment free of soil pests. With the onset of winter, the temperature decreases, as does the activity of PHB-destroying bacteria. Therefore, less insecticide is released. This is very ideal because, at the same time, pest activity is low. On the contrary, in spring, when the activity of pests is higher, more insecticide is released, and thus the plant is protected from pests.[111]

5.1.6.2 THE USE OF PHAS IN MEDICINE AND PHARMACEUTICALS

The biodegradability of PHAs and their compatibility with vital systems have caused this polymer to be used in the controlled release of drugs in pharmaceuticals and veterinary medicine.[35] These polymers are also used as biodegradable carriers for long-term doses of drugs in the body, surgical pins, suture threads, wound cleaning cotton, wound coverings, plates, bone replacement, blood vessel replacement, growth engine, and bone healing.

Bone and PHB have similar pyroelectric properties. Electrical stimulation strengthens bones. Therefore, when a fracture occurs in the bone, using a plate of PHB can hold the bone and stimulate its growth. On the other hand, because PHB is biodegradable, it is slowly absorbed, and resurgery to remove the graft is prevented.[12]

5.1.6.3 USE OF PHAS IN PACKAGING AND CONTAINER PRODUCTION

One of the simplest applications of P(HB-CO-HV) is in packaging, preparation of food containers, soft drink bottles, plastic films, and all kinds of bags. By using this polymer in the mentioned cases, the accumulation of waste in the environment is prevented.[12,41,71,121]

5.1.6.3.1 Application of PHAs

- Packaging films, bags, and containers.
- Biodegradable carriers for long-term use of drugs, insecticides, herbicides, and chemical fertilizers.
- Disposable items such as razors, diapers, and feminine hygiene products.
- Raw materials for compounds with optical properties (chiral).
- Bandages, threads, and surgical gloves.
- Covering wounds.
- Bone plates and replacements prostheses.
- Stimulating the growth and healing of bones with pyroelectric properties.
- Replacement of blood vessels.

5.2 METHODS

In this work, in connection with the paths of biological synthesis (biosynthesis pathways), the physical properties, recycling processes, and potential applications of degradable plastics obtained from PHAs were discussed. An anaerobic-aerobic activated sludge system was used in order to perform the necessary experiments. The action of sludge adaptation and the action of microorganisms to produce polymers in a sequencing batch reactor (SBR) system were investigated. Enrichment of microorganisms from polymers was carried out in a noncontinuous system (BATCH). After setting up the aforementioned systems and adapting them to the desired conditions, the effect of changing different parameters in each of the systems on the amount and type of polymer production was investigated. An outline of the system, which includes an anaerobic tank, an aerobic reactor, and a sedimentation tank, is shown in Figure 5.1. In this system, the incoming effluent is continuously injected into the anaerobic tank with a uniform flow rate.[131–149]

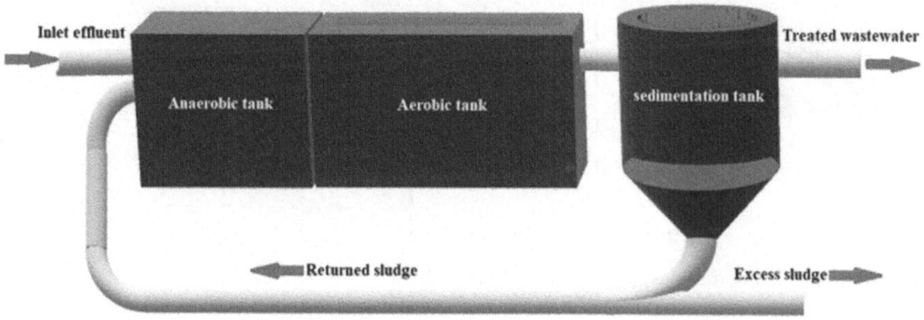

FIGURE 5.1 Polymer production system includes an anaerobic tank, an aerobic reactor, and a sedimentation tank.

The schematic of the polymer production system is shown in Figures 5.2 and 5.3. All the steps are precisely controlled by an electrical system. In this system, the intensity of oxidation and reduction reactions in aerobic and anaerobic stages will be measured and controlled by an oxidation–reduction potential (ORP) controller. In the field method, the data are received from pressure gauges and flowmeters with remote reading. The SBR system used in this work is shown in Figure 5.3. The SBR system was equipped with remote sensing (RS) and the Internet of Things (IoT).

The data intercommunication facilities were installed in the polymer production system. These facilities included an anaerobic tank, an aerobic reactor, a sedimentation tank, and parameter monitoring equipment. They were installed for measuring the PHA and reference parameters in the polymer production system. These facilities detected the PHA and Reference values from RS and IoT.[150–169]

During the process, the system was monitored continuously. Various parameters, including MLSS, MLVSS, COD, 3HB, and 3HV, were measured. In this system, all stages of sampling and addition of food were done automatically using an electronic control system. In order to control the pH, a pH meter controller with an accuracy of 0.1 unit was used. It was installed online in the system (Figure 5.3).

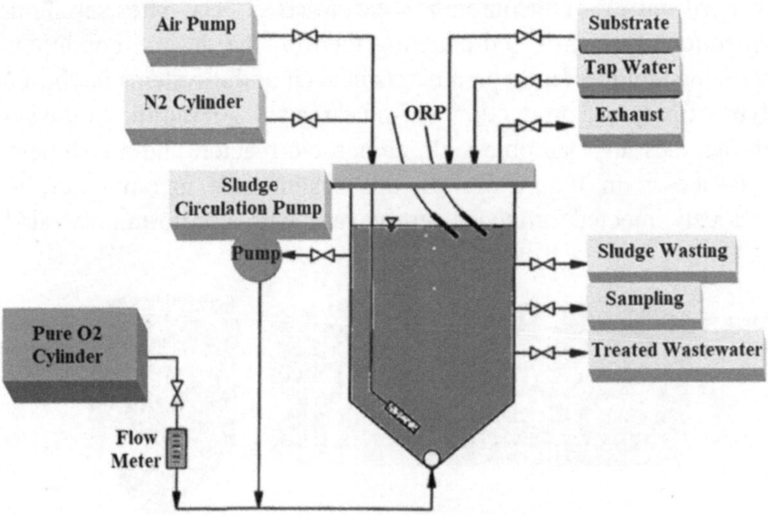

FIGURE 5.2 Schematic of styrene-butadiene rubber (SBR) system.

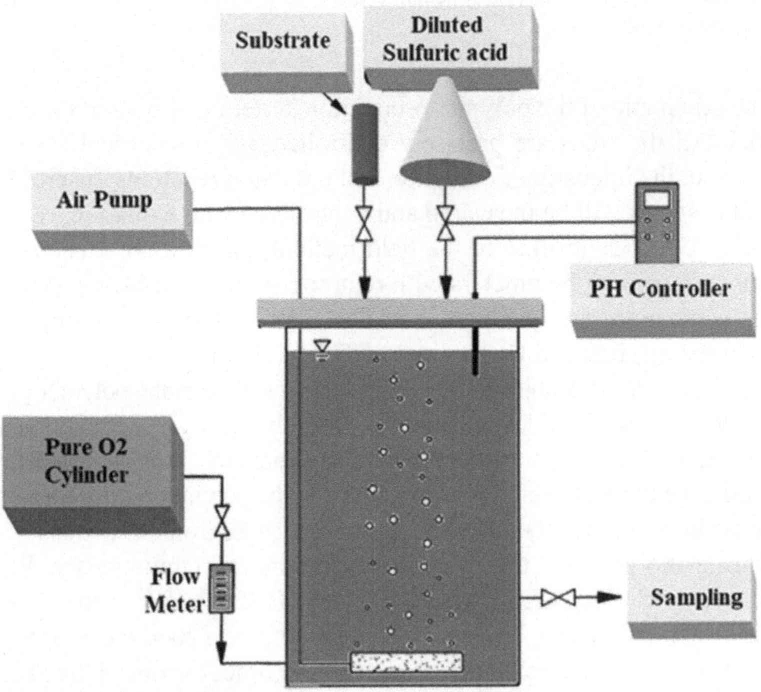

FIGURE 5.3 Styrene-butadiene rubber (SBR) for polymer production.

5.2.1 ANAEROBIC-AEROBIC ACTIVATED SLUDGE SYSTEM

The behavior of wastewater fluid flow in anaerobic-aerobic activated sludge system is a combination of the diffusing and remixing processes. In this work, various parameters, including MLSS, MLVSS, COD, 3HB, and 3HV, were investigated.

CO_2 is released, and O_2 is utilized by microorganisms as a result of their activity. The formulation of the process (eqs 5.2–5.12) in the anaerobic-aerobic activated sludge system and aeration tank is as following[23–25]:

$$V = LO/BV \tag{5.2}$$

$$BV = MLSS \times MLVSS \tag{5.3}$$

$$R_t = H / VS \tag{5.4}$$

$$R_t = V / Q$$

In the case of low *F/M*, microorganisms feed on organic material in wastewater or other microorganisms (Figure 5.1).

In order to save balancing conditions, we need to have high return sludge and high *MLSS*.

On the basis of European standards, this happened for days without rain.[28–31] The ratio of *RS* per *max* influent becomes 100% and can be achieved from the flowing relations:

$$RS / Q_{max} = [MLSS / (SMLSS - MLSS)] \times 100 \tag{5.5}$$

There was a problem in the treatment process when the sludge volume index (S.V.I.) reached more than 200. In this condition, sedimentation was failed. By decreasing S.V.I., aeration time was decreased.

5.2.1.1 SLUDGE VOLUME INDEX (S.V.I.)

The amount of surplus sludge that must be removed from the settling tank was related to the biochemical oxygen demand »BOD5« of the influent entrance to the aeration tank and settling tank output.

5.2.1.1.1 Sludge Age

$$\text{Sludge age} = (MLSS) \times (V) / (SS_e \times Q + SLS) \tag{5.6}$$

- Oxygenation Calculation:
  A portion of the absorbed pure oxygen is spent on energy consumption and multiplying of bacteria. The other part of O_2 is spent on the oxidation of organic carbonate and organic nitrogenous material.[165]

$$Q_V = A_Y B_V + B\,(MLSS) + 3 - 4\,(ON) \qquad (5.7)$$

- Process Design

$$O_C = [C_S / A(F)\,(C_S - C_X)] \times [.5YB_V + .1MLSS + 3.4ON] \qquad (5.8)$$

- Design Criteria:
  Influent production per person = 200 (L/day)
  Max influent factor = 1.71 (per 14 h)
  $BV = .5(KG.BOD5/M3DAY)$
  $BV / MLVSS = F/M = .15$ and
  $BOD5 = 60gr / p - d = .06kg / p - d$
  Wastewater temperature = 20°C

In this work, pure oxygen was introduced into the wastewater by a 150-bar pressure, 40-L-volume pure O_2 cylinder. By submerged porous diffusers and by air nozzles, pure O_2 was entered into the aeration tank. Then microorganisms that received pure O_2 grew quickly.[170–179]

5.2.1.1.2 Research Approach

Pollution decreasing to 20 (mg/L) for BOD5

Influent pollution calculation

- Fixed population = 700
- Office workers = 500
- Total population = 1200

$$Q_m^d = (700)\,(.2) + 500\,(.05) = 165\ (m^3/day)$$

Max influent per hour:

$Q_m^d = 165/14 = 12\ (\mathrm{m^3/h}) = 3.3\ (\mathrm{L/s})$

Total amount of influent pollution:

.06(1200) = 72(kg/p-day)

Average amount of influent pollution:

72/165 = .45 (kg/m³) = 450 (mg/L)

Due to pollution-detention curves, the decrease in influent pollution is related to detention time.

$R_t = V/Q$ = Tank volume/influent flow rate

$Rt + 200/12 = 17$(h)

Pollution decreasing percent = .35

450 (1 − 0.35) = 293 (mg/L)

Average amount of influent pollution:

(Per 24 h) =.293 (165) = 48.5 (kg BOD/day)

Oxygen Required Calculation Figure (eq 5.3):

$C_S = 9.17$ and $A = .9$

and

$Y = .925$ and $F = 0.85$

and

$MLSS = 3.3$ (kg/m³)

and

3.4(ON) = .23

$$O_C = [C_S/A(F)(C_S - C_X)] \times [.5(Y)B_V + .1(MLSS) + 3.4ON] \quad (5.9)$$

$$O_C = [9.17/.9(.85)(9.17 - 1.5)] \times [.5(.925)\ .5 + .1(3.3) + 3.4(ON)] \quad (5.10)$$

$$O_C = 1.2366\ (\text{kgO}_2/\text{m}^3.\text{day}) \quad (5.11)$$

$$\Sigma O_{C=}\ [200\ (1.2366)]/24 = 10.3\ (\text{kgO}_2/\text{h}) \quad (5.12)$$

5.3 RESULTS AND DISCUSSION

In this work, pure oxygen was introduced into wastewater by a 40-L pure O_2 cylinder. By submerged porous diffusers and air nozzles, pure O_2 was entered into the aeration tank. Microorganisms that received pure O_2 grew quickly. The important aim of this work was polymer production (Tables 5.1–5.60).

The miscible wastewater interpenetration happened when they moved themselves in the separated pipes toward the common joint and pipe in the aeration tank. Fluid parameters, including velocity, pressure, temperature, and other properties in the pipes, were homogeneous. The changes in the behavior of wastewater fluid flow and their effect on pure oxygen penetration in wastewater flow were investigated. A comparison between laboratory observations on fluid streamlines in pipes for three cases, including laminar flow, transient flow, and turbulent flow, showed little error in the (prediction of fluid flow software) software results.

Aeration Tank Volume = 200 (m^3)

Detention time:

$R_t = V/Q = 200/12 = 17(\text{h})$

Theoretical required volume:

$V = Lo/BV = 72/.5 = 144\ \text{m}^3$

Theoretical detention time:

$R_t = V/Q = 144/12 = 12\ (\text{h})$

5.3.1 PURE O_2 INJECTION IN WASTEWATER FOR PRODUCTION OF PHA

In this work, HB and HV units had photoactive carbon, and both had D (−) spatial form. This property of polymer was used for the chemical synthesis of photoactive substances that had biological activity. This polymer was used for the production of PHA.

Pure O_2 injection in wastewater for the production of PHA in the present work is shown in Figures 5.4–5.13.

5.3.2 DATA INTERCOMMUNICATION AND REGRESSION

The data intercommunication was carried out for the polymer production system, including the anaerobic tank, aerobic reactor, sedimentation tank, and monitoring equipment.

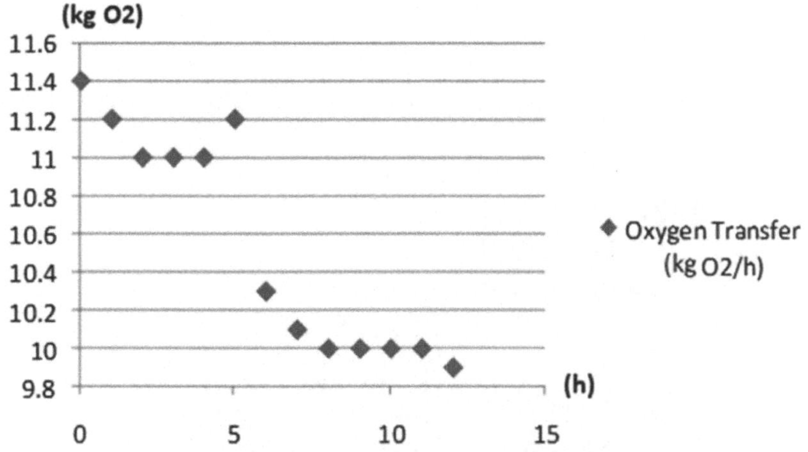

FIGURE 5.4 Pure oxygen required for the production of PHA.

FIGURE 5.5 Pure oxygen introduced in wastewater for polymer production.

FIGURE 5.6 Pure O_2 gas station in conventional activated sludge system.

FIGURE 5.7 Gas station for pure O_2 injection into wastewater.

FIGURE 5.8 Activated sludge system and aeration basin.

FIGURE 5.9 Sedimentation tank in activated sludge system.

FIGURE 5.10 Conventional activated sludge system.

FIGURE 5.11 Deep aeration in conventional activated sludge system.

FIGURE 5.12 Conventional activated sludge lagoon.

FIGURE 5.13 Aerobic and anaerobic tank; gas station for pure oxygen diffusion.

These data were used for regression analysis. The data were detected by networked sensors, which were installed for measuring the PHA and reference parameters in the polymer production system. The PHA and Reference values were also intercommunicated by RS and IoT. The PHA data and Reference values were transmitted to the control unit of the SBR system (Table 5.1).[180–197]

TABLE 5.1 Production of PHA Using Microorganisms.

Bacteria	Carbon source	PHA(%)	Reference
Gener alkali	Glucanate 50%	72.1	83
Gener alkali	Glucanate 50%	90.2	83
Gener alkali	Glucanate 50%	74.2	83
Gener alkali	Glucanate 50%	68.6	99.8
Gener alkali	Glucanate 50%	54.8	89
Gener alkali	Glucanate 50%	70.1	90
Gener alkali	Glucanate 50%	77.5	80

5.3.3 REGRESSION ANALYSIS

In regression analysis, the regression lines reflected the overall movement path of scattered points in the nominal coordinate system (Figures 5.14–5.27). The regression lines showed the intensity and weakness and the type of correlation between the variables, including the data of PHA (13) and Reference values (Tables 5.2–5.60).

$$PHA = f(\text{Reference}) \tag{5.13}$$

PHA—Dependent Variable: PHA (%)

H—Independent Variable: Reference

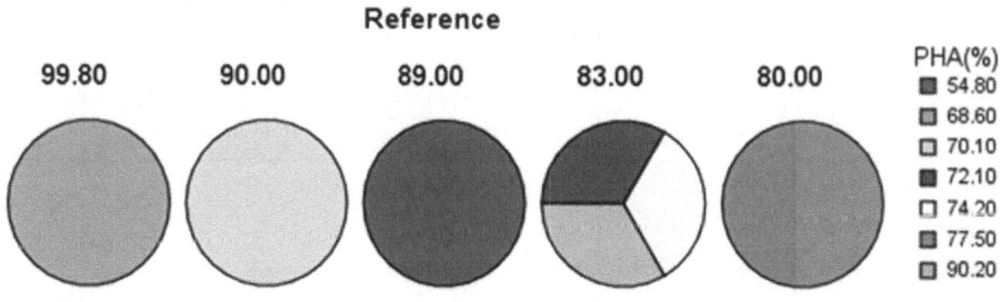

FIGURE 5.14 Processing summary.

TABLE 5.2 Model Description.

Model name		**MOD_1**
Dependent variable	1	PHA(%)
Equation	1	Linear
	2	Logarithmic
	3	Inverse
	4	Quadratic
	5	Cubic
	6	Compound[a]
	7	Power[a]
	8	S[a]
	9	Growth[a]
	10	Exponential[a]
	11	Logistic[a]
Independent variable		Reference
Constant		Included
Variable whose values label observations in plots		Unspecified
Tolerance for entering terms in equations		.0001

[a]The model requires all nonmissing values to be positive.

TABLE 5.3 Case Processing Summary.

	N
Total cases	7
Excluded cases[a]	0
Forecasted cases	0
Newly created cases	0

[a]Cases with a missing value in any variable are excluded from the analysis.

TABLE 5.4 Variables Processing Summary.

		Variables	
		Dependent	**Independent**
		PHA(%)	**Reference**
Number of positive values		7	7
Number of zeros		0	0
Number of negative values		0	0
Number of missing values	User-missing	0	0
	System-missing	0	0

TABLE 5.5 Model Summary and Parameter Estimates. Dependent Variable: PHA (%).

Equation	Model summary					Parameter estimates			
	R square	F	df1	df2	Sig.	Constant	b1	b2	b3
Linear	.224	1.444	1	5	.283	137.080	−.744		
Logarithmic	.237	1.549	1	5	.268	378.146	−68.508		
Inverse	.249	1.654	1	5	.255	−.030	6267.149		
Quadratic	.392	1.292	2	4	.369	1017.173	−20.420	.109	
Cubic	.396	1.309	2	4	.365	732.059	−10.734	.000	.000
Compound	.211	1.341	1	5	.299	174.129	.990		
Power	.225	1.449	1	5	.283	4813.987	−.943		
S	.238	1.558	1	5	.267	3.273	86.489		
Growth	.211	1.341	1	5	.299	5.160	−.010		
Exponential	.211	1.341	1	5	.299	174.129	−.010		
Logistic	.211	1.341	1	5	.299	.006	1.010		

The independent variable is Reference.

Due to the parameter limitations, the power function had a suitable correlation on the scatter diagram. The best-fit curve was used for PHA based on the regression model shown in Table 5.5. The percentage of PHA varied with the Reference value (eq 5.14), as shown in Figure 5.15.

$$f(x) = y = cx^{p} \tag{5.14}$$

$$y = 4813.987\times^{-0.943}$$

where:

y—PHA.
x—Reference value.

c and p are parameters determined by the linear least-squares method using data from the 10 parameters tested.

TABLE 5.6 Chi-Square Test for Dependent Variables.

	Observed N	Expected N	Residual
54.80	1	1.0	.0
68.60	1	1.0	.0
70.10	1	1.0	.0
72.10	1	1.0	.0
74.20	1	1.0	.0
77.50	1	1.0	.0
90.20	1	1.0	.0
Total	7		

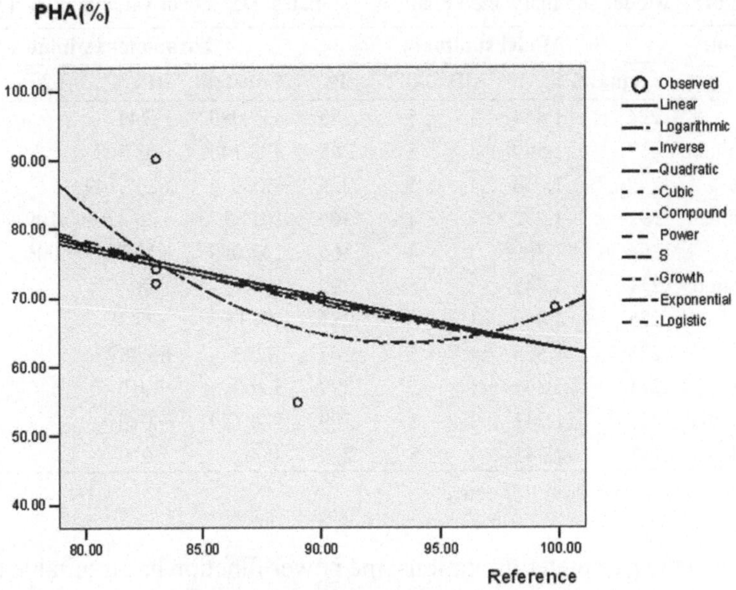

FIGURE 5.15 Curve fit for dependent and independent variables.

TABLE 5.7 Chi-Square Test for Independent Variables.

	Observed N	Expected N	Residual
80.00	1	1.4	−.4
83.00	3	1.4	1.6
89.00	1	1.4	−.4
90.00	1	1.4	−.4
99.80	1	1.4	−.4
Total	7		

TABLE 5.8 Chi-Square Test Statistics.

	PHA(%)	Reference
Chi-square[a,b]	.000	2.286
Df	6	4
Asymp. Sig.	1.000	.683

[a]seven cells (100.0%) have expected frequencies less than 5. The minimum expected cell frequency is 1.0.

[b]five cells (100.0%) have expected frequencies less than 5. The minimum expected cell frequency is 1.4.

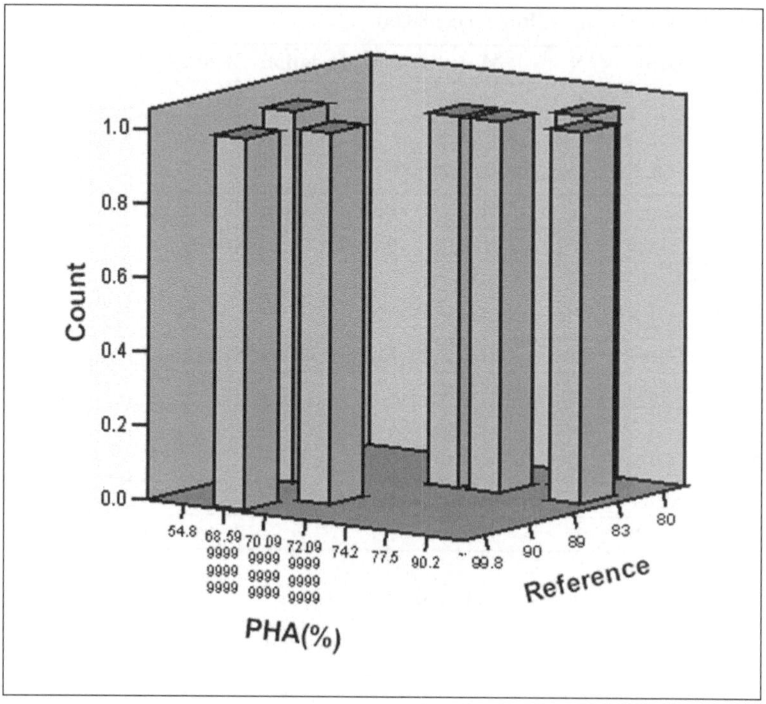

FIGURE 5.16 X graph for dependent and independent variables.

TABLE 5.9 Percent of Dependent Variable Values.

		Reference				
		80.00	**83.00**	**89.00**	**90.00**	**99.80**
PHA(%)	54.80			1		
	68.60					1
	70.10				1	
	72.10		1			
	74.20		1			
	77.50	1				
	90.20		1			

TABLE 5.10 Case Processing Summary.

	Cases					
	Included		**Excluded**		**Total**	
	N	**Percent**	**N**	**Percent**	**N**	**Percent**
PHA(%) * Reference	7	100.0%	0	.0%	7	100.0%

TABLE 5.11 OLAP Cubes; Reference: 80.00.

	Sum	N	Mean	Std. deviation	% of total sum	% of total N
PHA(%)	77.50	1	77.5000	.	15.3%	14.3%

TABLE 5.12 OLAP Cubes; Reference: 83.00.

	Sum	N	Mean	Std. deviation	% of total sum	% of total N
PHA(%)	236.50	3	78.8333	9.89966	46.6%	42.9%

TABLE 5.13 OLAP Cubes; Reference: 89.00.

	Sum	N	Mean	Std. deviation	% of total sum	% of total N
PHA(%)	54.80	1	54.8000	.	10.8%	14.3%

TABLE 5.14 OLAP Cubes; Reference: 90.00.

	Sum	N	Mean	Std. deviation	% of total sum	% of total N
PHA(%)	70.10	1	70.1000	.	13.8%	14.3%

TABLE 5.15 OLAP Cubes; Reference: 99.80.

	Sum	N	Mean	Std. deviation	% of total sum	% of total N
PHA(%)	68.60	1	68.6000	.	13.5%	14.3%

TABLE 5.16 OLAP Cubes; Reference: Total.

	Sum	N	Mean	Std. deviation	% of total sum	% of total N
PHA(%)	507.50	7	72.5000	10.61131	100.0%	100.0%

TABLE 5.17 Model Description.

Model name		MOD_1
Dependent variable	1	PHA(%)
Equation	1	Linear
Independent variable		Reference
Constant		Included
Variable whose values label observations in plots		Unspecified

TABLE 5.18 Case Processing Summary.

	N
Total cases	7
Excluded cases[a]	0
Forecasted cases	0
Newly created cases	0

[a]Cases with a missing value in any variable are excluded from the analysis.

TABLE 5.19 Variable Processing Summary.

		Variables	
		Dependent	Independent
		PHA(%)	Reference
Number of positive values		7	7
Number of zeros		0	0
Number of negative values		0	0
Number of missing values	User-missing	0	0
	System-missing	0	0

TABLE 5.20 Model Summary and Parameter Estimates; Dependent Variable: PHA (%).

Equation	Model summary					Parameter estimates	
	R square	F	df1	df2	Sig.	Constant	b1
Linear	.224	1.444	1	5	.283	137.080	−.744

The independent variable is Reference.

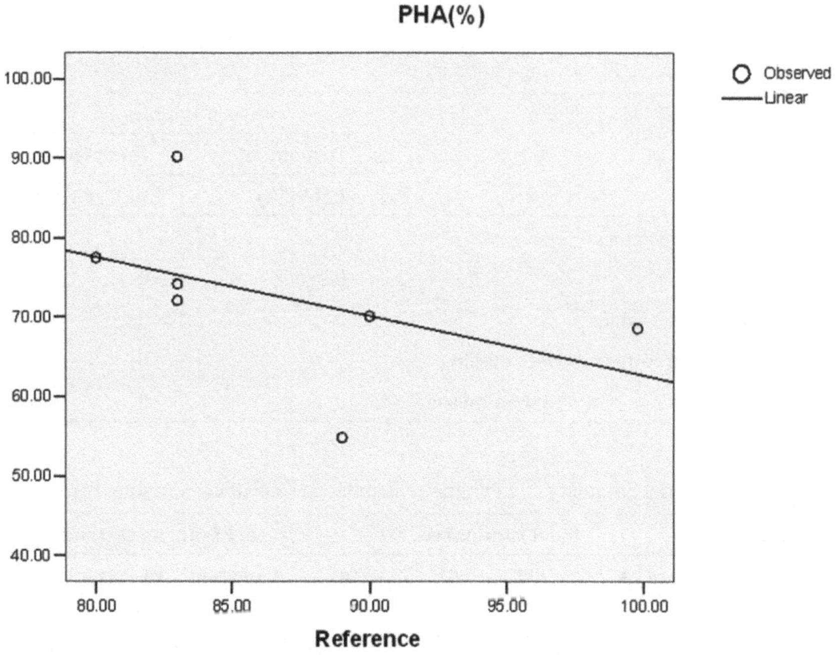

FIGURE 5.17 Curve fit; linear function.

TABLE 5.21 Curve Fit; Model Description.

Model name		MOD_3
Dependent variable	1	PHA(%)
Equation	1	Quadratic
Independent variable		Reference
Constant		Included
Variable whose values label observations in plots		Unspecified
Tolerance for entering terms in equations		.0001

TABLE 5.22 Case Processing Summary.

	N
Total cases	7
Excluded cases[a]	0
Forecasted cases	0
Newly created cases	0

[a]Cases with a missing value in any variable are excluded from the analysis.

TABLE 5.23 Variable Processing Summary.

		Variables	
		Dependent	Independent
		PHA(%)	Reference
Number of positive values		7	7
Number of zeros		0	0
Number of negative values		0	0
Number of missing values	User-missing	0	0
	System-missing	0	0

TABLE 5.24 Model Summary and Parameter Estimates; Dependent Variable: PHA (%).

Equation	Model summary					Parameter estimates		
	R square	F	df1	df2	Sig.	Constant	b1	b2
Quadratic	.392	1.292	2	4	.369	1017.173	−20.420	.109

The independent variable is Reference.

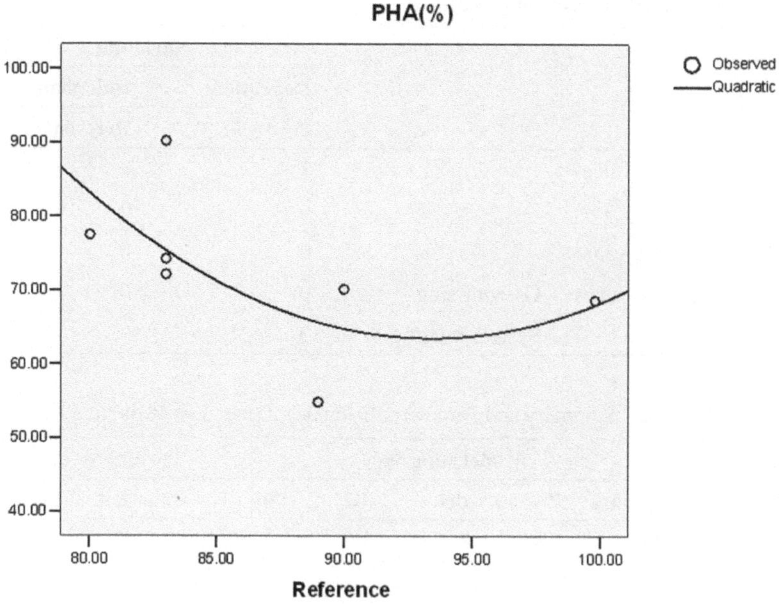

FIGURE 5.18 Curve fit; quadratic function.

TABLE 5.25 Curve Fit; Model Description.

Model name		**MOD_4**
Dependent variable	1	PHA(%)
Equation	1	Compound[a]
Independent variable		Reference
Constant		Included
Variable whose values label observations in plots		Unspecified

[a]The model requires all nonmissing values to be positive.

TABLE 5.26 Case Processing Summary.

	N
Total cases	7
Excluded cases[a]	0
Forecasted cases	0
Newly created cases	0

[a]Cases with a missing value in any variable are excluded from the analysis.

TABLE 5.27 Variable Processing Summary.

		Variables	
		Dependent	Independent
		PHA(%)	Reference
Number of positive values		7	7
Number of zeros		0	0
Number of negative values		0	0
Number of missing values	User-missing	0	0
	System-missing	0	0

TABLE 5.28 Model Summary and Parameter Estimates; Dependent Variable: PHA (%).

Equation	Model summary					Parameter estimates	
	R square	F	df1	df2	Sig.	Constant	b1
Compound	.211	1.341	1	5	.299	174.129	.990

The independent variable is Reference.

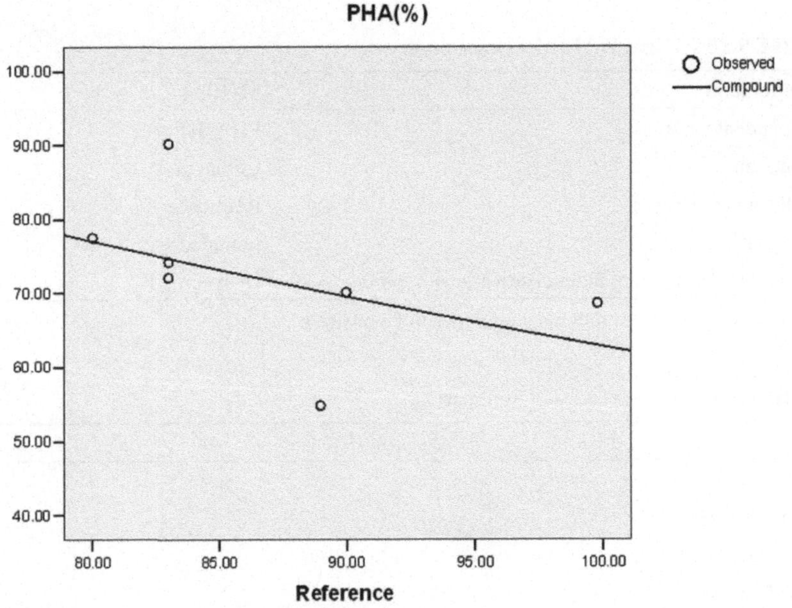

FIGURE 5.19 Curve fit; compound function.

TABLE 5.29 Curve Fit; Model Description.

Model name		MOD_5
Dependent variable	1	PHA(%)
Equation	1	Growth [a]
Independent variable		Reference
Constant		Included
Variable whose values label observations in plots		Unspecified

[a]The model requires all nonmissing values to be positive.

TABLE 5.30 Case Processing Summary.

	N
Total cases	7
Excluded cases[a]	0
Forecasted cases	0
Newly created cases	0

[a]Cases with a missing value in any variable are excluded from the analysis.

TABLE 5.31 Variable Processing Summary.

		Variables	
		Dependent	Independent
		PHA(%)	Reference
Number of positive values		7	7
Number of zeros		0	0
Number of negative values		0	0
Number of missing values	User-missing	0	0
	System-missing	0	0

TABLE 5.32 Model Summary and Parameter Estimates; Dependent Variable, PHA (%).

Equation	Model summary					Parameter estimates	
	R square	F	df1	df2	Sig.	Constant	b1
Growth	.211	1.341	1	5	.299	5.160	−.010

The independent variable is Reference.

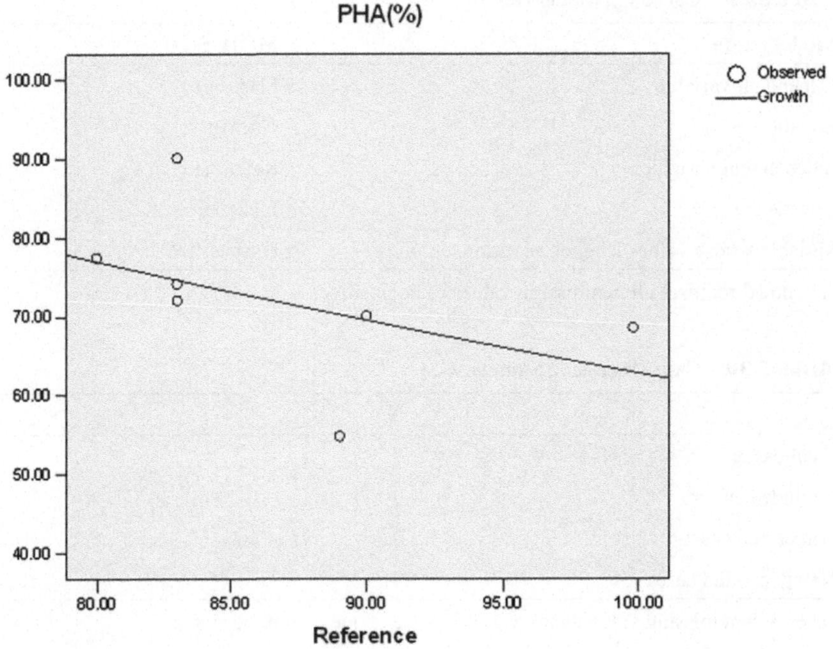

FIGURE 5.20 Curve fit; growth function.

TABLE 5.33 Case Processing Summary.

Model name		**MOD_6**
Dependent variable	1	PHA(%)
Equation	1	Logarithmic
Independent variable		Reference
Constant		Included
Variable whose values label observations in plots		Unspecified

TABLE 5.34 Case Processing Summary.

	N
Total cases	7
Excluded cases[a]	0
Forecasted cases	0
Newly created cases	0

[a]Cases with a missing value in any variable are excluded from the analysis.

TABLE 5.35 Variable Processing Summary.

		Variables	
		Dependent	**Independent**
		PHA(%)	**Reference**
Number of positive values		7	7
Number of zeros		0	0
Number of negative values		0	0
Number of missing values	User-missing	0	0
	System-missing	0	0

TABLE 5.36 Model Summary and Parameter Estimates; Dependent Variable: PHA (%).

Equation	**Model summary**					**Parameter estimates**	
	R square	**F**	**df1**	**df2**	**Sig.**	**Constant**	**b1**
Logarithmic	.237	1.549	1	5	.268	378.146	−68.508

The independent variable is Reference.

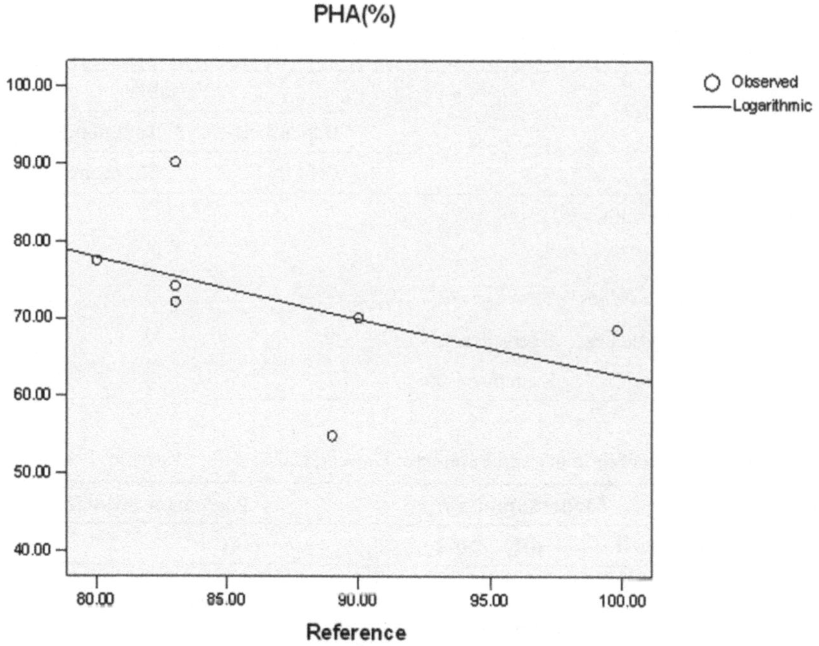

FIGURE 5.21 Curve fit; logarithmic function.

TABLE 5.37 Curve Fit; Model Description.

Model name		**MOD_7**
Dependent variable	1	PHA(%)
Equation	1	Cubic
Independent variable		Reference
Constant		Included
Variable whose values label observations in plots		Unspecified
Tolerance for entering terms in equations		.0001

TABLE 5.38 Case Processing Summary.

	N
Total cases	7
Excluded cases[a]	0
Forecasted cases	0
Newly created cases	0

[a]Cases with a missing value in any variable are excluded from the analysis.

TABLE 5.39 Variable Processing Summary.

		Variables	
		Dependent	**Independent**
		PHA(%)	**Reference**
Number of positive values		7	7
Number of zeros		0	0
Number of negative values		0	0
Number of missing values	User-missing	0	0
	System-missing	0	0

TABLE 5.40 Model Summary and Parameter Estimates; Dependent Variable: PHA (%).

Equation	**Model summary**					**Parameter estimates**			
	R square	**F**	**df1**	**df2**	**Sig.**	**Constant**	**b1**	**b2**	**b3**
Cubic	.396	1.309	2	4	.365	732.059	−10.734	.000	.000

The independent variable is Reference.

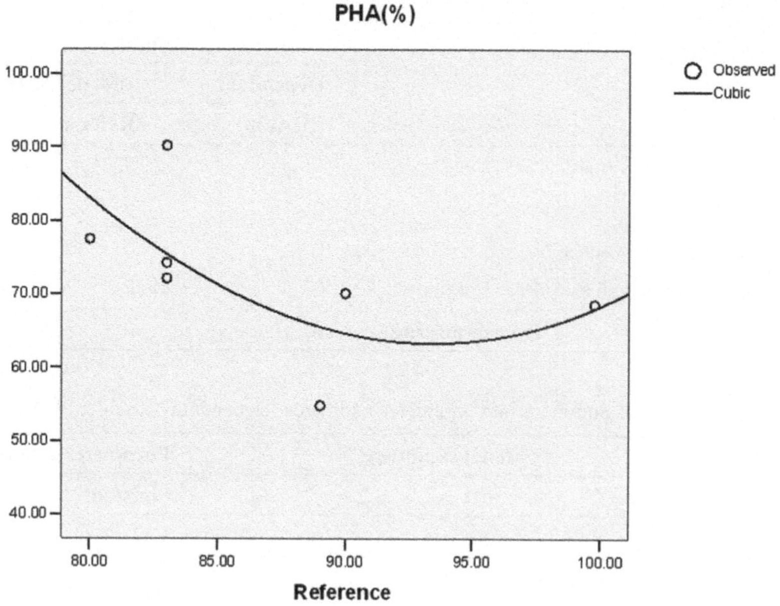

FIGURE 5.22 Curve fit; cubic function.

TABLE 5.41 Curve Fit; Model Description.

Model name		**MOD_8**
Dependent variable	1	PHA(%)
Equation	1	S[a]
Independent variable		Reference
Constant		Included
Variable whose values label observations in plots		Unspecified

[a]The model requires all nonmissing values to be positive.

TABLE 5.42 Case Processing Summary.

	N
Total cases	7
Excluded cases[a]	0
Forecasted cases	0
Newly created cases	0

[a]Cases with a missing value in any variable are excluded from the analysis.

TABLE 5.43 Variable Processing Summary.

		Variables	
		Dependent	Independent
		PHA(%)	Reference
Number of positive values		7	7
Number of zeros		0	0
Number of negative values		0	0
Number of missing values	User-missing	0	0
	System-missing	0	0

TABLE 5.44 Model Summary and Parameter Estimates; Dependent Variable: PHA (%).

Equation	Model summary					Parameter estimates	
	R square	F	df1	df2	Sig.	Constant	b1
S	.238	1.558	1	5	.267	3.273	86.489

The independent variable is Reference.

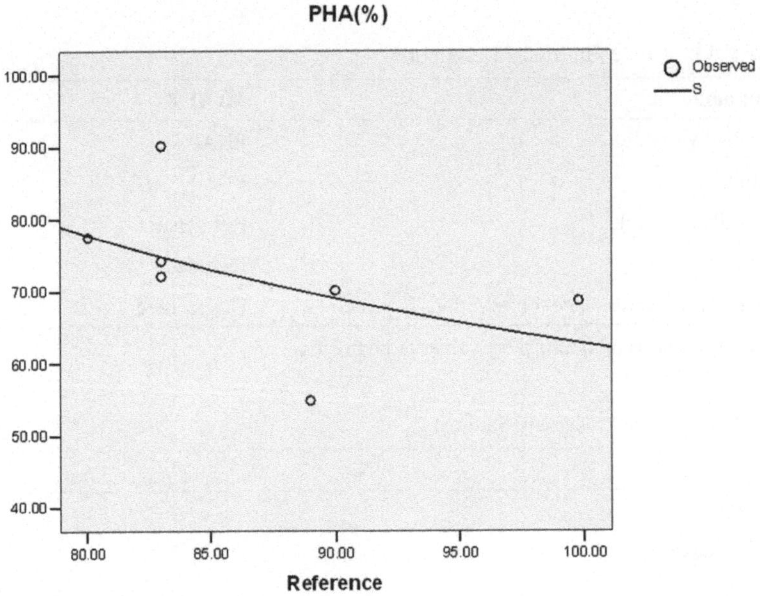

FIGURE 5.23 Curve fit; S function.

TABLE 5.45 Curve Fit; Model Description.

Model name		**MOD_9**
Dependent variable	1	PHA(%)
Equation	1	Exponential[a]
Independent variable		Reference
Constant		Included
Variable whose values label observations in plots		Unspecified

[a]The model requires all nonmissing values to be positive.

TABLE 5.46 Case Processing Summary.

	N
Total cases	7
Excluded cases[a]	0
Forecasted cases	0
Newly created cases	0

[a]Cases with a missing value in any variable are excluded from the analysis.

TABLE 5.47 Variable Processing Summary.

		Variables	
		Dependent	**Independent**
		PHA(%)	**Reference**
Number of positive values		7	7
Number of zeros		0	0
Number of negative values		0	0
Number of missing values	User-missing	0	0
	System-missing	0	0

TABLE 5.48 Model Summary and Parameter Estimates; Dependent Variable: PHA (%).

Equation	**Model summary**					**Parameter estimates**	
	R square	**F**	**df1**	**df2**	**Sig.**	**Constant**	**b1**
Exponential	.211	1.341	1	5	.299	174.129	−.010

The independent variable is Reference.

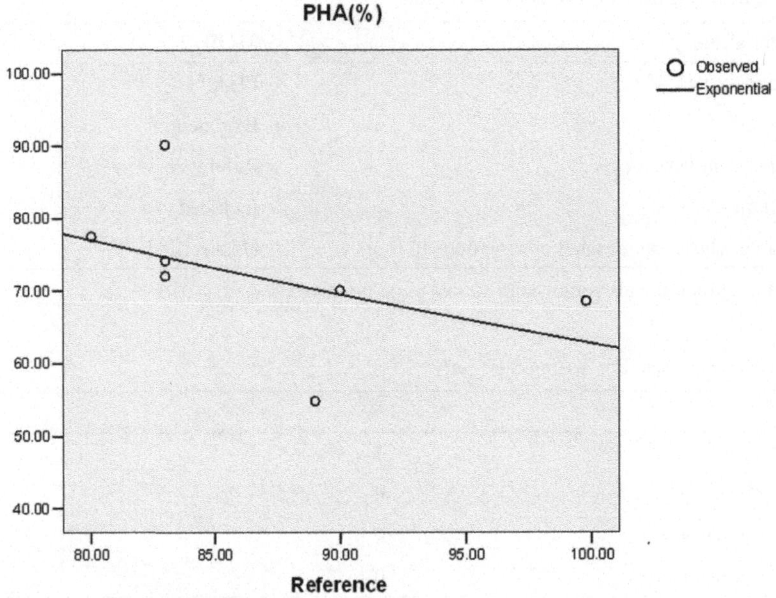

FIGURE 5.24 Curve fit; exponential function.

TABLE 5.49 Curve Fit; Model Description.

Model name		**MOD_10**
Dependent variable	1	PHA(%)
Equation	1	Inverse
Independent variable		Reference
Constant		Included
Variable whose values label observations in plots		Unspecified

TABLE 5.50 Case Processing Summary.

	N
Total cases	7
Excluded cases[a]	0
Forecasted cases	0
Newly created cases	0

[a]Cases with a missing value in any variable are excluded from the analysis.

TABLE 5.51 Variable Processing Summary.

		Variables	
		Dependent	**Independent**
		PHA(%)	**Reference**
Number of positive values		7	7
Number of zeros		0	0
Number of negative values		0	0
Number of missing values	User-missing	0	0
	System-missing	0	0

TABLE 5.52 Model Summary and Parameter Estimates; Dependent Variable: PHA (%).

Equation	**Model summary**					**Parameter estimates**	
	R square	**F**	**df1**	**df2**	**Sig.**	**Constant**	**b1**
Inverse	.249	1.654	1	5	.255	−.030	6267.149

The independent variable is Reference.

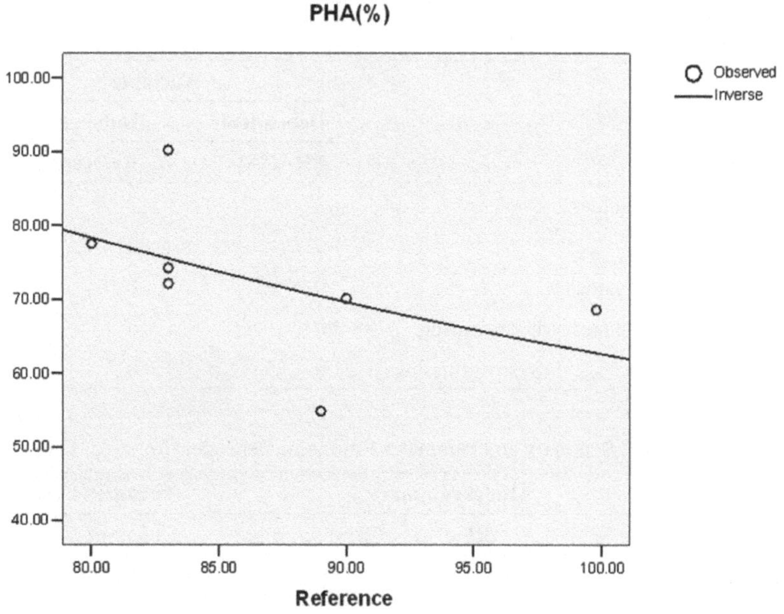

FIGURE 5.25 Curve fit; inverse function.

TABLE 5.53 Curve Fit; Model Description.

Model name		**MOD_11**
Dependent variable	1	PHA(%)
Equation	1	Power[a]
Independent variable		Reference
Constant		Included
Variable whose values label observations in plots		Unspecified

[a]The model requires all nonmissing values to be positive.

TABLE 5.54 Case Processing Summary.

	N
Total cases	7
Excluded cases[a]	0
Forecasted cases	0
Newly created cases	0

[a]Cases with a missing value in any variable are excluded from the analysis.

TABLE 5.55 Variable Processing Summary.

		Variables	
		Dependent	**Independent**
		PHA(%)	**Reference**
Number of positive values		7	7
Number of zeros		0	0
Number of negative values		0	0
Number of missing values	User-missing	0	0
	System-missing	0	0

TABLE 5.56 Model Summary and Parameter Estimates; Dependent Variable: PHA (%).

Equation	**Model summary**					**Parameter estimates**	
	R square	**F**	**df1**	**df2**	**Sig.**	**Constant**	**b1**
Power	.225	1.449	1	5	.283	4813.987	−.943

The independent variable is Reference.

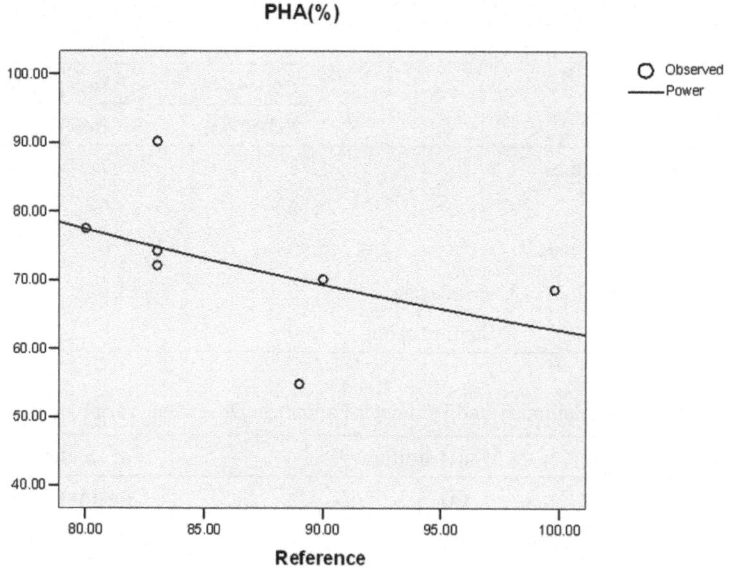

FIGURE 5.26 Curve fit; power function.

TABLE 5.57 Curve Fit; Model Description.

Model name		**MOD_12**
Dependent variable	1	PHA(%)
Equation	1	Logistic[a]
Independent variable		Reference
Constant		Included
Variable whose values label observations in plots		Unspecified

[a]The model requires all nonmissing values to be positive.

TABLE 5.58 Case Processing Summary.

	N
Total cases	7
Excluded cases[a]	0
Forecasted cases	0
Newly created cases	0

[a]Cases with a missing value in any variable are excluded from the analysis.

TABLE 5.59 Variable Processing Summary.

		Variables	
		Dependent	Independent
		PHA(%)	Reference
Number of positive values		7	7
Number of zeros		0	0
Number of negative values		0	0
Number of missing values	User-missing	0	0
	System-missing	0	0

TABLE 5.60 Model Summary and Parameter Estimates; Dependent Variable: PHA (%).

Equation	Model summary					Parameter estimates	
	R square	F	df1	df2	Sig.	Constant	b1
Logistic	.211	1.341	1	5	.299	.006	1.010

The independent variable is Reference.

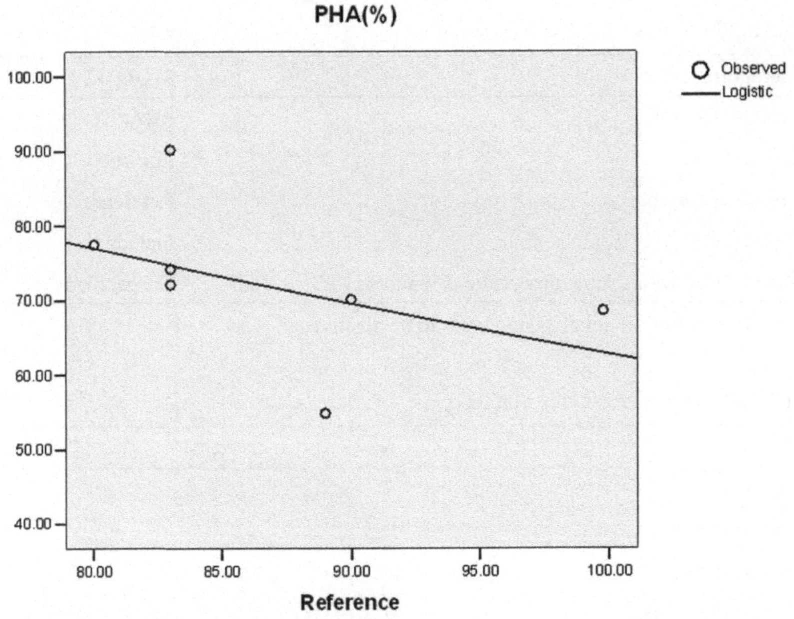

FIGURE 5.27 Curve fit; logistic function.

5.3.4 REGRESSION AND SCATTER DIAGRAM

In this work, there was a linear relationship between the variables of the research. The scatter diagram gave the basic idea. The regression equations were used to draw the regression line.

5.3.4.1 RELATIONSHIP BETWEEN PHA AND REFERENCE VALUE

Based on the main results of this work, for the production of PHA, there was a significant relationship between the PHA and the Reference value in the production process of PHA (eqs 5.13 and 5.14).

$$\text{PHA} = f(\text{Rerence}) \tag{5.13}$$

PHA—Dependent Variable: PHA (%)
H—Independent Variable: Reference

$$f(x) = y = cx^{\mathrm{p}} \tag{5.14}$$

$$y = 4813.987 \times^{-0.943}$$

where:
y—PHA.
x—Reference value.

The results of this work showed that there was a significant relationship between PHA and the Reference value. By regression analysis and through investigation of the scatter diagram and the curve fit due to regression analysis, PHA against the Reference value eq 5.13 showed a good correlation between dependent variable (PHA) and independent variable (Reference value) data (Figures 5.15–5.27).

5.4 CONCLUSION

The result of this work showed that urban wastewater can be used as a cheap source for the production of PHA. It can also be used for the production of bioplastics. These types of plastics have properties that are comparable to those of synthetic plastics. Among these properties, we can mention 100% water resistance and the low cost of its production compared to other degradable bioplastics.

With regard to the above, taking into account the problems related to the disposal of plastics used in product packaging as well as waste transportation and considering the costs related to the disposal of excess sludge from sewage treatment plants, the production of PHAs by using activated sludge and the determination of factors effective have been chosen as the subject of this work.

KEYWORDS

- **activated sludge**
- **anaerobic**
- **synthetic plastics**
- **aerobic**
- **networked sensors**

REFERENCES

1. Chua, A. S. M.; Takabatake, H.; Satoh, H.; Takashi Mino, T. Production of Polyhydroxyalkanoayes (PHA) by Activated Sludge Treating Municipal Wastewater: Effect of pH, Sludge Retention Time (SRT), and Acetate Concentration in Influent. *Water Res.* **2003,** *37*, 3602–3611.
2. Akiyama, M.; Taima, Y. M.; Doi, Y. Production of PHB by a Bacterium of the Genus alcaligeses is Utilizing Long Chain Fatty Acids. *Appl. Microbiol. Biotechnol.* **1992,** *37*, 689–701.
3. Akiyama, M.; Doi, Y. Production of PHB from Alkanedioic Acids and Hydroxylated Fatty Acids by Alcaligesesis sp. *Biotechnol. Lett.* **1993,** *15* (2), 163–168.
4. Anderson, A. J.; Dawes, E. A. Occurrence, Metabolism, Metabolic Role, and Industrial uses of Bacterial Polyhydroxyalkanoates. *Microbiol. Rev.* **1990,** *54* (4), 450–472.
5. Anderson, A. J.; Haywood, G. W.; Williams, D. R.; Dawes, E. A. The Production of Polyhydroxyalkanoates from Unrelated Carbon Sources. In *Novel Biodegradable Microbial Polymers*; Dawes, E. A., Ed.; Kluwer Academic Publishers: The Netherlands, 1990; pp 119–129.
6. Anderson, A. J.; Williams, D. R.; Taidi, B.; Dawes, E. A.; Ewing, D. F. Studies in Copolyester Synthesis by Rhodococcus Rubber and Factors Influencing the Molecular Mass of Polyhydroxybutyrate Accumulated by Methylobacterium Extorquens and Alcaligenes Eutrophus. *FEMS Microbiol. Rev.* **1992,** *130*, 93–102.
7. Ayorinde, F. O.; Saeed, K. A.; Eribo, E.; Morrow, A.; Collis, W. E.; Mclnnis, F.; Polack, S. K.; Eribo, B. Production of PHB from Saponified Veronica Galamesis Oil by Alcaligenesis eutrophus. *J. Indus. Microbiol. Biotechnol.* **1998,** *21*, 46–50.

8. Berger, E.; Ramsay, B. A.; Ramsay, J. A.; Chavarie, C. PHB Recovery by Hypochlorite Digestion of Non-PHB Biomass. *Biotech. Techniq.* **1989,** *3* (4), 227–232.
9. Booma, M.; Selke, S. E.; Giacin, J. R. Degradable Plastics. *J. Elastomers Plast.* **1994,** *26*, 104–142.
10. Bourque, B. O.; Andre, G.; Groleau, D. Production of Poly-hydroxybutyrate from Methanol: Characterization of a New Isolate of Methylobacteruim Extorquens. *Appl. Microbiol. Biotechnol.* **1992,** *37*, 7–12.
11. Bourque, D.; Pomerleau, Y.; Groleau, D. High Cell Density Production of Poly-hydroxdybutyrate (PHB) form Methanol by Methylobacteruim Extorquens: Production of High-Molecular-Mass PHB. *Appl. Microbiol. Biotechonol.* **1995,** *44*, 367–376.
12. Brandl, H.; Gross, R. A.; Lenz, R. W.; Fuller, R. C. Plastics form Bacteria and for Bacteria: Poly (hydroxybutyrate) as Natural, Biodegradable Polyesters. *Adv. Biochem. Eng.* **1990,** *41*, 77–93.
13. Brandl, H.; Bachofen, B.; Mayer, J.; Wintermantel, E. Degradation and Applications of Polyhydroxyalkanoates. *Can. J. Microbiol.* **1995,** *41* (suppl.1), 143–153.
14. Braunegg, G.; Sonnleitner, B.; Lafferty, R. M. A Rapid Gas Chromatographic Method for the Determination of PHB in Microbial Biomass. *Eur. J. Appl. Microb. Biotech.* **1978,** *6*, 905–910.
15. Braunegg, G.; Lefebvere, G.; Genser, K. F. Polyalkanoates, Biopolyester from Renewable Resources: Physiological and Engineering Aspects. *J. Biotechnol.* **1998,** *65*,127–161.
16. Byrom, D. Polymer Synthesis by Microorganisms: Technology and Economics. *Tibtech* **1987,** *5*, 246–250.
17. Byrom, D. Industrial Production of Copolymer from Alcaligenes eutrophus. In *Novel Biodegradable Microbial Polymers*; Dawes. E. A., Ed.; Kluwer Academic Pubishers: the Netherlands, 1990; pp 113–117.
18. Byrom, D. Production of Poly-hydroxybutyrate: Poly-hydroxyvalerate Copolymers. *FEMS Microbiol. Rev.* **1992,** *103*, 247–250.
19. Castella, J. M.; Urmenta, J.; Lafuente, R.; Navarrte, A.; Guerrero, R. Biodegradation of PHAs in Aerobic Sediments. *Inter. Biodeter. Biodegrad.* **1995,** 155–174.
20. Chen, Y.; Chen, J.; Yu, C.; Du, G.; Lun, S. Recovery of Poly-3-hydroxybutyrate from Alcaligenes Eutrophus by Surfactant-Chelate Aqueous System. *Process Biochem.* **1999,** *34* (2), 153–157.
21. Chen, G.; Page, W. J. The Effect of Substrate on the Molecular Weight of PHB Production by Azotobacter vinelandii UWD. *Biotechnol. Let.* **1994,** *16* (2), 155–160.
22. Cho, G. D.; Yoo, J. Y.; Oh, J. T.; Kim, W. S. Study on the Biosynthesis of PHB with Alcaligenesis latus. *Hwahak Konghak* **1997,** *35*, 412–418.
23. Chua, H.; Yu, P. H. F.; Ho, L. Y. Recovery of Biodegradable Polymers from Food-Processing Wastewater Activated Sludge System, *J. IES* **1997,** *37* (2), 9–13.
24. Chua, H.; Yu, P. H. F. Production of Biodegradable Plastics form Chemical Wastewater-A Novel Method to Reduce Excess Activated Sludge Generated from Industrial Wastewater Treatment. *Wat. Sci. Tech.* **1999,** *39* (10-11), 273–280.
25. Chuang, S. H.; Ouyang, C. F.; Yuang, H. C.; You, S. J. Evaluation of Phosphorus Removal in Anaerobic- Anoxic- Aerobic aystem- via Polyhydroxyalkanoates Measurements. *Water Sci. Tech.* **1998,** *38* (1), 107–114.
26. Chuang, S. H.; Ouyang, C. F.; Yuang, H. C.; You, S. J. Effects of SRT and DO on Nutrient Removal in a Combined A.S.-Biofilm Process. *Water Sci. Tech.* **1997,** *36* (12), 19–27.

27. Daniel, M.; Choi, J. H.; Kim, J. H.; Lebeault, J. M. Effect of Nutrient Deficiency on Accumulation and Relative Molecular Weight of PHB by Methylotrophic Bacterium Pseudomonas 135. *Appl. Microbiol. Biotechnol.* **1992,** *37*, 702–706.
28. Dawes, E. A.; Senior, P. J. The Role and Regulation of Energy Reserve Polymers in Micro-organisms. *Adv. Microb. Physiol.* **1973,** *10*, 135–266.
29. Dawes, E. A. Starvation, Survival and Energy Reserves. In *Bacteria in their Natural Environment*; Fletcher, M., Floodgate, G. D., Eds.; Academic Press: Florida, 1985; pp 43–79.
30. Dawes, E. A. Novel Microbial Polymers: An Introductory Overview. In *Novel Biodegradable Microbial Polymers*; Dawes, E. A., Ed.; Kluwer Academic Publishers: The Netherlands, 1990; pp 3–16.
31. De Koning, G. Physical Properties of Bacterial Poly(R)-3hydroxyalkanoates). *Can. J. Microbiol.* **1995,** *41* (suppl.1), 303–309.
32. Dionisi, D.; Majone, M.; Ramadori, R.; Beccari, M. The Storage of Acetate Under Anoxic Conditions. *Water Res.* **2001,** *35* (11), 2661–2668.
33. Doi, Y.; Kunioka, M.; Nakamura, Y.; Soga, K. Nuclear Magnetic Resonance Studies on Poly (hydroxybutyrate) and a Copolymer of Hydroxybutyrate and Hydroxyvalerate Isolated from Alcaligenes eutrophus H16. *Macromolecules* **1986,** *19*, 2860–2864.
34. Doi, Y.; Kunioka, M.; Nakamura, Y.; Soga, K. Biosynthesis of Copolyesters in Alcaligenes eutrophus H16 from 13C-lebeled Acetate and Propionate. *Macromolecules* **1987,** *20*, 2988–2991.
35. Doi, Y. In *Microbial Polyesters*; VCH Publishers, Inc.: New York, USA, 1990.
36. Doi, Y.; Segawa, A.; Nakamura, S.; Kunioka, M. Production of Biodegradable Copolymers by Alcaligenes eutrophus. In *Novel Biodegradable Microbial Polymers*; Dawes, E. A., Ed.; Kluwer Academic Pubishers: The Netherlands, 1990; pp 37–48.
37. Fidler, S.; Dennis, D. Polyhydroxyalkanoate Production in Recombinant Escherichia coli. *FEMS Microbiol. Rev.* **1992,** *103*, 231–236.
38. Fuchtenbusch, B.; Steinbuchel, A. Biosynthesis of PHAs from Low Rank Coal Liquefaction Products by Pseudomonas Oleovorance and Rhodococcus Rubber. *Appl. Microbiol. Biotechnol.* **1999,** *52*, 91–95.
39. Fuchtenbusch, B.; Wullbrandt, D.; Steinbuchel, A. Production of PHAs by Ralstonia eutroph and Pseudomonas Oleovorance from Oil Remaining from Biotechnological Rhamnose Production. *Appl. Microbiol. Biotechnol.* **2000,** *53*, 167–172.
40. Gorenflo, V.; Steinbuchel, A.; Marose, S.; Rieseberg, M.; Scheper, T. Quantification of Bacterial Polyhydroxyalkanoic Acids by Nile Red Staining. *Appl. Microbiol. Biotechnol.* **1999,** *51*, 765–772.
41. Griffin, G. J. L. In *Chemistry and Technology of Biodegradable Polymers*, 1st Ed.; Chapman & Hall, 1994.
42. Gross, R. A.; DeMello, C.; Lenz, R. W.; Brandl, H.; Fuller, C. Biosynthesis and Characterization of PHAs Produced by Pseudomnas Oleovorance. *Macromolecules* **1989,** *22*, 1106–1115.
43. Grothe, E.; Chisti, Y. PHB Thermoplastic Production by Alcaligenesis Latus Behavior of Fed-Batch Culture. *Bioprocess Eng.* **2000,** *22*, 441–449.
44. Hahn, S. K.; Chang, Y. L.; Kim, B. S.; Chang, H. N. Optimization of Microbial Poly (3-hydroxybutyrate) Recovery using Dispersions of Sodium Hypochlorite Solution and Chloroform. *Biotechnol. Bioeng.* **1994,** *44*, 250–262.

45. Harrison, S.; Dennis, J. S.; Chase, S. A. The Effect of Culture History on the Distruption of Alcaligenes eutrophus by High Pressure Homogenization. In *Separation for Biotechnology II*, 1st Ed.;Pyle, D. L., Ed.; Elsvier: London, 1990.
46. Hassan, Mohd A.; et al. The Production of Polyhydroxyalkanoate form Anaerobically Treated Palm Oil Mill Effluent by Rhodobacter sphaeroides. *J. Fermentation Bioeng.* **1997,** *83* (5), 485–488.
47. Haywood, G. W.; Anderson, A. J.; Dawes, E. A. A Survey of the Accumulation of Novel Polyhydroxyalkanoates by Bacteria. *Biotechnol. Lett.* **1989,** *11* (7), 471–476.
48. Haywood, G. W.; Anderson, A. J.; Williams, D. R.; Dawes, E. A. Accumulation of a Poly (hydroxyalkanoate) Copolymer Containing Primarily 3-Hydroxyvarate from Simple Carbohydrate Substrates by Rhodococcus sp. NCIMB 40126. *Int. J. Biol. Macromol.* **1991,** *13*, 83–88.
49. Hejazi, P.; Vasheghani-Farahani, E.; Yamini, Y. Supercritical Fluid Disruption of Ralstonia eutropha for Poly (hydroxybutirate) Recovery. *Biotechnol. Prog.* **2003,** *19*, 1519–1523.
50. Hezayen, F. F.; Rehm, B. H. A.; Eberhardt, R. Polymer Production by 2 Newly Isolation Extremely Halophilic Archaea: Application of a Novel Corrosion Resistance Bioreactor. *Appl. Microbiol. Biotechnol.* **2000,** *54* (3), 319–325.
51. Holmes, P. A. Application of PHB-a Microbially Produced Biodegradable Thermoplastics. *Phys. Technol.* **1985,** *16*, 32–36.
52. Hood, C. R.; Randall, A. A. A Biochemical Hypothesis Explaining the Response of Enhanced Biological Phosphorus Removal Biomss to Organic Substrates. *Water Res.* **2001,** *35* (11), 2758–2766.
53. Horib, K.; Marsudi, S.; Unno, H. Simulatanous Production of PHAs and Rhamnolipids by Pseudomonas aeroginosa. *Biotechnol. Bioeng.* **2002,** *78* (6), 699–707.
54. Hrabak, O. Industrial Production of Poly-/3-hydroxybutyrate. *FEMS Microbiol. Rev.* **1992,** *103*, 251–256.
55. Huang, J.; Shetty, A. S.; Wang, M. Biodegradable Plastics: A Review. *Adv. Polym. Technol.* **1990,** *10* (1), 23–30.
56. 1Ishihara, Y.; Shimizu, H.; Shioya, S. Mole Fraction of PHB in Fed-Batch Culture of A.eutrophus. *Biotechnol. Bioeng.* **1994,** *81* (5), 422–428.
57. Jan, S.; Roblot, C.; Courtois, B.; Babotin, J. N.; Seguin, J. P. H NMR Spectroscopic Determination of Poly-hydroxybutyrate Extracted from Microbial Biomass. *Enz. Microb. Technol.* **1996,** *18*, 195–201.
58. Yu, J.; Si, Y.; Keung, W.; Wong, R. Kinetics Modeling of Inhibition and Utilization of Mixed Volatile Fatty Acids in the Formation of Polyhydroxyalkanoates by Ralstonia eutropha. *Process Biochem.* **2002,** *37*, 731–738.
59. Jung, K.; Hazenberg, W.; Prieto, M. W. Two Stage Continuous Process Development for the Production of Medium Chain Length PHAs. *Biotechnol. Bioeng.* **2001,** *72* (1), 19–24.
60. Juttner, P. R.; Lafferty, R. M.; Knackmuss, H. Y. A Simple Method for the Determination of Poly-hydroxybutyric Acid in Microbial Biomass. *Europ. J. Appl. Microbiol.* **1975,** *1*, 233–237.
61. Kato, N.; Konishi, H.; Shimo, M.; Sakazawa, C. Production of PHB Trimmer by Bacillus megaterium B124. *J. Fermen. Bioeng.* **1992,** *73* (3), 246–247.
62. Kellerhals, M. B.; Hazenberg, W.; Witholt, B. High Cell Density Fermentation of Pseudomonas oleovorance for the Production of mcl-PHA in Two Liquid Phase Media. *Enz. Microbial. Technol.* **1999,** *24*, 111–116.

63. Khatipov, E.; Miyake, M.; Miyake, J.; Asads, Y. Accumulation of PHB by Rhodobacter sphearoides on Variouse Carbon Source and Nitrogen Substrates. *FEMS Microbiol. Lett.* **1998,** *1652*, 39–45.
64. Khosravi-Darani, K.; Vasheghani-Farahani, E.; Shojaosadati, S. A.; Yamini, Y. Effect of Process Variables on Supercritical Fluid Disruption of Ralstonia eutropha Cells for poly (R- hydroxybutirate) Recovery. *Biotechnol. Prog.* **2004,** *20*, 1757–1765.
65. Kim, B. S.; Lee, S. C.; Lee, S. Y.; Chang, H. N. Production of PHB by Fed-batch Culture of Ralstonia eutropha with Glucose Concentration Control. *Biotecnol. Bioeng.* **1994,** *43*, 892–898.
66. Kimura, H.; Yoshida, Y.; Doi, Y. Production of PHB by Pseudomonas Acidovorance. *Biotech. Lett.* **1992,** *14* (6), 149–158.
67. Kulaev, I. S.; Vagabov, V. M. Polyphosphate Metabolism in Microorganisms. *Adv. Microbial. Physiol.* **1983,** *24*, 83–171.
68. Kumagai, Y. Enzymatic Degradation of Binary Blends of Microbial poly (3-Hydroxy butyrate) with Enzymatically Active Polymers. *Polym. Degrad. Stab.* **1992,** *37*, 253–256.
69. Lafferty, R. M.; Korsatko, B.; Korsatko, W. Microbial Production of Poly-hydroxybutyric Acid. In *Biotechnology*; Rehm, H. J., Reed, G., Eds.; VCH Publishers: New York, 1988; pp 135–176.
70. Law, J.; Slepecky, R. Assay of Poly-hydroxybutyric Acid. *J. Bacteriol.* **1961,** *82*, 33–36.
71. Lee, S. Y. Review Bacterial Polyhydroxyalkanoates. *Biotechnol. Bioeng.* 1996, *49*, 1–14.
72. Lee, S.; Yu, J. Production of Biodegradable Thermoplastics from Municipal Sludge by a Two-stage Bioprocess. *Resour. Conserv. Recycl.* **1997,** *19*, 151–164.
73. Lee, S. Y.; Choi, J. Production and Degradation of Polyhydroxyalkanoates in Waste Environment. *Waste Manag.* **1999,** *19* (2), 133–139.
74. Lemos, P. C.; et al. Effect of Carbon Source on the Formation of Polyhydroxyalkanoates (PHA) by a Phosphate-Accumulation Mixed Culture. *Enzyme Microb. Techonol.* **1998,** *22*, 662–671.
75. Libergesel, M.; Husted, E.; Timm, A.; Steinbuchel, A.; Fuller, R. C.; Len, Z. R. W.; Schlegel, H. G. Formation of PHA by Phototrophic and Chemulithotrophic Bacteria. *Arch. Microbial.* **1991,** *155*, 415–421.
76. Ling, Y.; Wong, H. H.; Thomas, C. J.; Williams, D. R. G.; Middelberg, A. P. J. Pilot-Scale Extraction of PHB from Recombinant E. coli by Homogenization and Centrifugation. *Biosepatation* **1997,** *7*, 9–15.
77. Linko, S.; Vaheri, H.; Seppala, J. Production of PHB by Alcaligenesis eutrophus on Different Carbon Sources. *Appl. Microbiol. Biotechnol.* **1993,** *39*, 11–15.
78. Madigan, M. T.; Martinko, J. M.; Parker, J. In *Brock Biology of Microorganisms*; Princeton Hall: New Jersey, 1997.
79. Maranogoni, C.; Furigo, A.; Glancia, M. F. Production of PHB by Ralstonia eutropha in Whey and Inverted Suger with Propionic Acid Feeding. *Process Biochem.* **2002,** *38*, 137–141.
80. Masow, T.; Babel, W. Calorimetrically Recognized Maximum Yield of PHB Continuously Synthesized from Toxic Substrate. *J. Biotechnol.* **2000,** *77*, 247–253.
81. McCool, G. J.; Fernandez, T.; Li, N.; Cannon, M. Polyhydroxyaklanoate Inclusion- Body Growth and Prokiferation in Bacillus megaterium. *FEMS Microbial. Lett.* **1996,** *138*, 41–48.
82. Mergeay, M.; Houba, C.; Gerits, J. Extra Chromosomal Inheritance Controlling Resistance to Cadmium and Zinc Ions: Evidence from Curing in a Pseudomonas. *Arch. Int. Physiol. Biochem.* **1978,** *86*, 440–441.

83. Mino, T.; Kawakmi, T.; Matsuo, T. Location of Phosphorus in Activated Sludge and Fraction of Intracellular Polyphosphate in Biological Phosphorus Removal Process. *Wat. Sci. Technol.* **1985a,** *17* (2/3), 93–106.
84. Mino, T.; Kawakami, T.; Matsuo, T. Behavior of Intracellular Polyphosphate in Biological Phosphorus Removal Process. *Wat. Sci. Technol.* **1985b,** *17* (11,12), 11–21.
85. Ohi, K.; Takaida, N.; Komemushi, S.; Okazaki, M.; Muira, Y. A New Species of Hydrogen- Oxidizing Bacterium. *Appl. Microbiol.* **1979,** *25*, 53–58.
86. Page, W. J. Production of PHB by Azotobacter vinelandii UWD During Growth on Molasses and Other Complex Carbon Sources. *Appl. Microbiol. Biotechnol.* **1989,** *31*, 329–333.
87. Page, W. J. Production of Polyhydroxyalkanoates by Azotobacter vinelandii Strain UWD in Beet Molasses Culture. *FEMS Microbiol. Rev.* **1992,** *103*, 149–158.
88. Page, W. J.; Cornish, A. Growth of Azotobacter vinelandii in Fish Peptone Medium and Simplified Extraction of PHB. *Appl. Environ. Microbiol.* **1993,** 4236–4244.
89. Page, W. J. Bacterial Polyhydroxyalkanoates, Natural Biodegradable Plastics with a Great Future. *Can J. Microbiol.* **1995,** *41* (suppl. 1), 1–3.
90. Page, W. J.; Manchak, J. The Role of Oxidation of Short-Chain Alkanoates in Polyhydroxyalkanoate Copolymer Synthesis in Azotobacter vinelandii UWD. *Can. J. Microbiol.* **1995,** *41* (suppl.1), 106–115.
91. Pelissero, A. Update on Biodegradable Plastics Materials. *Imballaggio* **1987,** *38*, 54.
92. Pfeffer, J. T. Recycling. In *Solid Waste Management Engineering*, 1992; pp 72–84.
93. Preiss, J. Bacterial Glycogen Synthesis and its Regulation. *Ann. Rev. Microbiol.* **1984,** *38*, 419–458.
94. Preusting, H.; Kingama, J.; Witholt, B. Physiology and Polyester Formation of Pseudomonas Oleovorance in Continuous Two Liquid Phase Cultures. *Enzy. Microbial. Technol.* **1991,** *13*, 770–780.
95. Preusting, H.; Hazenberg, W. Continuous Production of PHB by Pesudomonas Oleovorance in a High Cell; Density, Two Liquid Phases Chemostat. *Enz. Microbial. Technol.* **1993,** *15*, 311–316.
96. Ramsay, J. A.; Berger, E.; Voyer, C.; Chavarie, C.; Ramsay, B. A. Extraction of Poly-3-hydroxybutyrate Using Chlorinated Solvents. *Biotechnol. Techninq.* **1990,** *8* (8), 589–594.
97. Ramsay, J. A.; Berger, E.; Ramsay, B. A.; Chavarie, C. Recovery of Poly-3- hydroxyalkanoic Acid Granules by a Surfactant-Hypochlorite Treatment. *Biotechnol. Techniq.* **1994,** *9* (10), 709–712.
98. Ramsay, J. A.; Berger, E.; Voyer, R.; Chavarie, C.; Ramsay, B. A. Extraction of Poly-3-hydroxybutyrate Using Chlorinated Solvents. *Biotech. Techniq.* **1994,** *8* (8), 589–594.
99. Ramsay, J. A.; Berger, E.; Ramsay, B. A.; Chavarie, C. Recovery of Poly-hydroxyalkanoic Acid Granules by a Surfactant-Hypochlorite Treatment. *Biotechnol. Techniq.* **1990,** *4* (4), 221–226.
100. Renner, G.; Haage, G.; Braunegg, G. Production of Short-Side-Chain Polyhydroxyalkanoates by Various Bacteria from the rRNA Superfamily III. *Appl. Microbiol. Biotechnol.* **1996,** *46*, 268–272.
101. Resch, S.; Gruber, K.; Wanner, G.; Slater, S.; Dennnis, D.; Lubitz, W. Aqueous Release and Purification of Poly (Hydroxybutyrate) from Escherichia coli. *J. Biotechnol.* **1998,** *65*, 173–182.
102. Roh, K. S.; Yeom, S. H.; Yoo, Y. J. The Effects of Sodium Bisulfate in Extraction of PHB by Hypochlorite. *Biotechnol. Techniq.* **1995,** *4* (4), 221–226.

103. Roy, R. K. In *A Primer on the Tguchi Method*, 1st Ed.; Van Nostrand Reinhold,1990,
104. Saad, B.; Neuenschwander, P.; Uhlschmid, G. K.; Suter, U. W. New Versatile Elastomeric Degradable Polymer Materials for Medicine. *Int. J. Biol. Macromol.* **2001,** 158–163.
105. Lee, S.; Yu, J. Production of Biodegradable Thermoplastics from Municipal Sludge by a Two Stage Bioprocess. *Resour. Conserv. Recycl.* **1997,** *19*, 151–164.
106. Sasikala, C. H.; Ramana, C. H. V. Biodegradable Polyesters. In *Advances in Applied Microbiology*; Neidleman, S. L., Laskin, A. I., Eds.; Academic Press: California, 1996; vol 42, pp 97–218.
107. Satoh, H.; Mino, T.; Matsuo, T. Uptake of Organic Substrates and Accumulation of Polyhydroxyalkanoates Liked with Glycolysis of Intracellular Carbohydrates Under Anaerobic Conditions in the Biological Excess Phosphorus Removal Processes. *Wat. Sci. Tech.* **1992,** *26* (5-6), 933–942.
108. Satoh, H.; Iwamoto, Y.; Mino, T.; Matsuo, T. In *Activated Sludge as a Possible Source of Biodegradable Plastic*, Proceedings, Water Quality International 1998; In Book 3 Wastewater Treatments; IAWQ 19th Biennial International Conference: Vancouver, BC, Canada, June 21–26, 1998.
109. Schwien, U.; Schmidt, E. Improved Degradation of Monochlorophenol by Constructed Strain. *Appl. Environ. Microbial.* **1982,** *44*, 33–39.
110. Senior, P. J.; Beech, G. A.; Ritchie, G. A. F.; Dawes, E. A. The Role of Oxygen Limitation in the Formation of Poly-hydroxybutyrate During Batch and Continuous Culture of Azotobacter beijerinckii. *Biochem. J.* **1972,** *128*, 1193–1201.
111. Shahhosseini, S. A Fed-Batch Model for PHA Production Using Alcaligenes eutrophus, M.S. Thesis, Department of Chemical Engineering, The University of Queensland, Australia, 1994.
112. Shimao M. Biodegradation of Plastics. *Curr. Opin. Biotechnol.* **2001,** *12* (3), 242–247.
113. Shimizu, H.; Tamira, S.; Shioya, S.; Suga, K. Kinetics Study of PHB Productionjn and its Molecular Weight Distribution Control in Fed-Batch Culture of A. eutrophus. *J. Ferment. Bioeng.* **1993,** *76* (6), 465–469.
114. Shimizu, H.; Tamira, S.; Shioya, S.; Suga, K. Kinetics Study of PHB Productionjn and its Molecular Weight Distribution Control in Fed-Batch Culture of A. eutrophus. *J. Fermenta. Bioeng.* **1993,** *76* (6), 465–469.
115. Shimizu, H.; Tamura, S.; Ishihara, Y.; Shioya, S.; Suga, K. Control of Molecular Weight Distribution and Mole Fraction in Poly (D 3hydroxyalkanoates) (PHA) Production by Alcaligenes eutrohpus. In *Biodegradable Plastics and Polymers*; Doi, Y., Fukuda, K., Eds.; Elsevier Science B.V.: New York, 1994; pp 365–372.
116. Shimizu, H.; Kozaki, Y.; Kodama, H.; Shioya, H. Maximum Production Strategy for Biodegradable Copolymer PHBV in Fed-Batch Culture of Alcaligenesis eutrophus. *Biotechnol. Bioeng.* **1998,** *62* (5), 518–525.
117. Smolders, G. J. F.; van der Meij, J.; van Loosdrecht, M. C. M. Model of the Anaerobic Metabolism of the Biological Phosphorus Removal Process: Stochiometry and pH Influence. *Biotechnol. Bioeng.* **1994,** *43*, 461–470.
118. Sonneleitner, B.; Heinzle, G.; Braunegg, G.; Lafferty, R. M. Formal Kinetics of PHB Production in Alcaligenesis eutrophus H16 and Mycoplana rubra to the Dissolved Oxygen Tension in Ammonium-Limited Batch Cultures. *Eur. J. Appl. Microbiol.* **1979,** *7*, 1–10.
119. Steinbuchel, A.; Pieper, U. Production of a Copolyester of 3-hydroxybutyric Acid and 3-hydroxyvaleric Acid from Single Unrelated Carbon Sources by a Mutant of Alcaligenenes eutrophus. *Appl. Microbiol. Biotechnol.* **1992,** *37*, 1–6.

120. Steinbuchel, A. PHB and Other Polyhydroxyalkanoic Acids. In *Biotechnology*; Rehm, H. J., Reed, G., Eds.; VCH: New York, 1996; pp 403–464.
121. Steinbuchel. A.; Fuchtenbusch, B. Bacterial and Other Biological Systems for Polyester Production. *TIBTECH* **1998,** *16*, 419–427.
122. Suzuki, T.; Yamane, T.; Shimizu, S. Mass Production of Poly-hydroxybutyric Acid by Fully Automatic Fed-Batch Culture of Methylotroph. *Appl. Microbiol. Biotechnol.* **1986,** *23*, 322–329.
123. Suzuki, T.; Yamane, T.; Shimizu, S. Mass Production of PHB by Fully Automatic Fed-Batch Culture of Methylotroph. *Appl. Microbiol. Biotechnol.* **1989,** *23*, 322–329.
124. Takabatake, H.; Satoh, H.; Mino, T.; Matsuo, T. Recovery of Biodegradable Plastics from Activated Sludge Process. *Wat. Sci. Tech.* **2000,** *42* (3-4), 351–356.
125. Tsuchikura, K. BIOPOL Properties and Processing. In *Biodegradable Plastics and Polymers*; Doi, Y., Fukuda, K., Eds.; Elsevier Science B.V.: New York, 1994; pp 362–364.
126. Van Groenestijn, J. W.; Deinema, M. H.; Zehnder, A. J. B. ATP Production from Polyphosphate in Acinetobacter Strain 210A. *Arch. Microbial.* **1987,** *148*, 14–19.
127. Varma, A. K.; Peck, H. D. Utilization of Short and Long-Chain Polyphosphate as Energy Sources for the Anaerobic Growth of Bacteria. *FEMS Microbial. Lett.* **1983,** *16*, 281–285.
128. Yamane, T. Yield of Poly-D (-)-3hydroxybutyrate from Various Carbon Sources: A Theoretical Study. *Biotechnol. Bioeng.* **1993,** *41* (1), 165–170.
129. Yamane, T.; Fukunage, M.; Lee, Y. W. Increase PHB Productivity by High-Cell- Density Fed–Batch Culture of Alcaligenes latus, a Growth-Associated PHB Producer. *Biotechnol. Bioeng.* **1996,** *50*, 197–202.
130. Yu, P. H. F.; Chua, H.; Huang, A. L.; Lo, W. H.; Ho, K. P. Transformation of Industrial Food Wastes into Polyhydroxyalkanoates. *Water Sci. Tech.* **1999,** *40* (1), 365–370.
131. Kevin A. Internet of Things. *RFID J.* June 22, 2009.
132. Parker, D. S.; Merrill, M. S. Oxygen and Air Activated Sludge: Another View. *J. Water Pollut. Control Fed.* **1976,** *48* (11), 2511–2528. http://www.jstor.org/stable/25040054
133. Asli, K. H.; Aliyev, S. A. O.; Thomas, S.; Deepu, A. In *Handbook of Research for Fluid and Solid Mechanics: Theory, Simulation, and Experiment*, 1st Ed.; Apple Academic Press, 2017. https://doi.org/10.1201/9781315365701.
134. Asli, K. H.; Naghiyev, F. B. O.; Haghi, R. K.; Asli, H. H. In *Advances in Control and Automation of Water Systems*, 1st Ed.; Apple Academic Press, 2012. https://doi.org/10.1201/b13114.
135. Hariri, A. K. Nonlinear Dynamics Modeling by Geospatial Information System (GIS). *Mapping Geospatial Inf. J. Guilan* **2019,** *3* (1). www.mpogl.ir.
136. https://en.wikipedia.org/wiki/Artificial_intelligence.
137. Kelly, K. New Rules for the New Economy, first published in 1998 by Viking Penguin, 1998.
138. https://fa.wikipedia.org/wiki/ Geographic Information System.
139. Wylie, E. B.; Streeter, V. L. In *Fluid Transients*; Feb Press: Ann Arbor, MI, 1983, corrected copy: pp 166–171, 1982.
140. Lee, T. S.; Pejovic, S. Air Influence on Similarity of Hydraulic Transients and Vibrations. *ASME J. Fluid Eng.* **1996,** *4*, 706–709.
141. Yoon, S. H.; Park C. S. Objective Building Energy Performance Benchmarking Using Data Envelopment Analysis and Monte Carlo Sampling. *Sustainability* **2017,** *9*, 780. DOI: 10.3390/su9050780.

142. Hariri, A. H.; Nazari, S. Water Age and Leakage in Reservoirs; Some Computational Aspects and Practical Hints. *Larhyss J.* **2021,** *48*, 151–167. http://larhyss.net/ojs/index.php/larhyss/article/view/808/807.
143. Hariri, A. H.; Hozori, A. Non-Revenue Water (NRW) and 3d Hierarchical Model for Landslide. *Larhyss J.* **2021,** *48*, 189–210. http://larhyss.net/ojs/index.php/larhyss/article/view/810/810.
144. Kensek, K. Buildings. **2015,** *5*, 899–916. DOI: 10.3390 /buildings5030899.
145. Wang, S. K. In *Handbook of Air-Conditioning and Refrigeration*; McGraw-Hill, Inc., 1994. ISBN 0-07-068138-4.
146. https://fa.wikipedia.org/wiki/ Internet of Things _IoT.
147. Ferras, D.; Manso, A. P. A.; Schleiss, A. J.; Covas, D. I. C. One-dimensional Fluid-Structure Interaction Models in Pressurized Fluid-Filled Pipes: A Review. *Appl. Sci.* **2018,** *8*, 1844–1856. https://doi.org/10.3390/app8101844
148. Vardy, A. E.; Brown, J. M. B.; He, S.; Ariyaratne, C.; Gorji, S. Applicability of Frozen-Viscosity Models of Unsteady Wall Shear Stress. *J. Hydraul. Eng.* **2015,** *141*, 1–13. https://doi.org/10.1061/(ASCE)HY.1943-7900.0000930
149. Asli, K. H.; Aliyev, S. A. O.; Thomas, S.; Deepu, A. In *Handbook of Research for Fluid and Solid Mechanics: Theory, Simulation, and Experiment*, 1st Ed.; Apple Academic Press, 2017. https://doi.org/10.1201/9781315365701
150. Asli, K. H.; Naghiyev, F. B. O.; Haghi, R. K.; Asli, H. H. In *Advances in Control and Automation of Water Systems*, 1st Ed.; Apple Academic Press, 2012. https://doi.org/10.1201/b13114
151. Lema, M.; Peña, F. L.; Buchlin, J. M.; Rambaud, P.; Steelant, J. Analysis of Fluid Hammer Occurrence with Phase Change and Column Separation Due to Fast Valve Opening by Means of Flow Visualization. *Exp. Therm. Fluid Sci.* **2016,** *79*, 143–153. DOI10.1016/j.expthermflusci.2016.07.008
152. Zecchin, A. C.; Maier, H. R.; Simpson, A. R.; Leonard, M.; Nixon, J. B. Parametric Study for an Ant Algorithm Applied to Water Distribution System Optimization. IEEE Trans. Evolut. Comput. **2005,** *9*, 175–191. DOI: 10.1109/TEVC.2005.844168
153. Andrade, D. M.; de Freitas Rachid, F. B. A Versatile Friction Model for Newtonian Liquids Flowing Under Unsteady Regimes in Pipes. *Meccanica* **2022,** *57*, 43–72. https://doi.org/10.1007/s11012-021-01458-5
154. Sarkardeh, H.; Zarrati, A. R.; Jabbari, E.; Marosi, M. Numerical Simulation and Analysis of Flow in a Reservoir in the Presence of Vortex. *J. Eng. Appl. Comput. Fluid Mech.* **2014,** *8* (4), 598–608. DOI: 10.1080/19942060.2014.11083310
155. Zhao, L.; Yang, Y.; Wang, T.; Han, W.; Wu, R.; Wang, P.; Wang, Q.; Zhou, L. An Experimental Study on the Water Hammer with Cavity Collapse under Multiple Interruptions. *Water* **2020,** *12*, 2566. https://doi.org/10.3390/w12092566
156. Duan, H. F.; Pan, B.; Wang, M.; Chen, L.; Zheng, F.; Zhang, Y. State-of-the-Art Review on the Transient Flow Modeling and Utilization for Urban Water Supply System (UWSS) Management. *J. Water Supply Res. Trans.* **2020,** *69*, 858–893. https://doi.org/10.2166/aqua.2020.048
157. Zhou, L.; Liu, D.; Karney, B. Investigation on Hydraulic Transients of Two Entrapped Air Pockets in a Water Pipeline. *J. Hydraul. Eng.* **2013**. DOI: 10.1061/(ASCE)HY.1943-7900.0000750
158. Möller, G.; Detert, M.; Boes, R. M. Vortex-Induced Air Entrainment Rates at Intakes. *J. Hydraul. Eng.* **2015**. DOI: 10.1061/(ASCE)HY.1943-7900.0001036

159. Hariri, A. K.; Nagiyev, F. B.; Haghi, A. K. Interpenetration of Two Fluids at Parallel Between Plates and Turbulent Moving in Pipe. In *Computational Methods in Applied Science and Engineering*, Chapter 7; Nova Science Publisher: USA, 2009; pp 115–128. https://www.novapublishers.com/catalog/product_ info.php?products_id=10681.
160. Hariri, A. K.; Nagiyev, F. B.; Haghi, A. K. Some Aspects of Physical and Numerical Modeling of Water Hammer in Pipelines. *Int. J. Nonlinear Dyn. Chaos Eng. Syst.* **2009**. ISSN: 0924-090X (print version) ISSN: 1573-269X (electronic version) Journal no. 11071; Springer: USA, http://nody.edmgr.com/
161. Hariri, A. H.; Arabani, M.; Golpour, Y. Reclaimed Asphalt Pavement (RAP) Based on a Geospatial Information System (GIS). *Slovak J. Civil Eng.*; Slovak University of Technology in Bratislava: Slovak **2020,** *28* (2), 36–42. https://doi.org/10.2478/sjce-2020-0013.
162. Hariri, A. H. Investigation of the Factors Affecting Pedestrian Accidents in Urban Roundabouts. *Trans. Civil Environ. Eng.* **2022,** *8* Article ID: 2226, 1–4, Special Issue: NCTT 2021. https://doi.org/10.52547/crpase.8.1.2255.
163. Hariri, A. H. M. Arabani, Analysis of Strain and Failure of Asphalt Pavement, Computational Research Progress in Applied Science & Engineering. *Trans. Civil Environ. Eng.* **2022,** *8*, 1–11, Article ID: 2250. https://doi.org/10.52547/crpase.8.1.2250.
164. Lee, T. S.; Pejovic, S. Air Influence on Similarity of Hydraulic Transients and Vibrations. *ASME J. Fluid Eng.* **1996,** *118* (4), 706–709.
165. Monzavi, M. T. Wastewater Treatment, Tehran, Iran, 1991; pp 7–162.
166. Filion, Y.; Karney, B. W. In *A Numerical Exploration of Transient Decay Mechanisms in Water Distribution Systems*, Proceedings of the ASCE Environmental Water Resources Institute Conference; American Society of Civil Engineers: Roanoke, Virginia, 2002.
167. Shirzad, S.; Hassan, M. M.; Aguirre, M. A.; Mohammad, L. N.; Daly, W. H. Evaluation of Sunflower Oil as a Rejuvenator and its Microencapsulation as a Healing Agent. *J. Mater. Civil Eng.* **2016,***28* (11), 04016116. https://ascelibrary.org/doi/abs/10.1061/%28 ASCE%29MT.1943-5533.0001625.
168. Chen, M.; Leng, B.; Wu, S.; Sang, Y. Physical, Chemical, and Rheological Properties of Waste Edible Vegetable Oil Rejuvenated Asphalt Binders. *J. Constr. Build. Mater.* **2014,** *66* (15 Sep), 286–298. https://doi.org/10.1016/j.conbuildmat.2014.05.033.
169. Fini, E. H.; Kalberer, E. W.; Shahbazi, A.; Basti, M.; You, Z.; Ozer, H. Chemical Characterization of Bio-binder from Swine Manure: Sustainable Modifier for Asphalt Binder. *J. Mater. Civil Eng.* **2011,** *23* (11), 1506–1513. https://ascelibrary.org/doi/abs/ 10.1061/%28ASCE%29MT.1943-5533.0000237.
170. Hassan, M.; Essam, T.; Yassin, A. S.; Salama, A. Optimization of Rhamnolipid Production by Biodegrading Bacterial Isolates Using Plackett–Burman Design. *Int. J. Biol. Macromol.* **2016,** *82* (4), 573–579. DOI: 10.1016/j.ijbiomac.2015.09.057.
171. https://en.wikipedia.org/wiki/Computational_physics
172. NZS 3106, Design of Concrete Structures for the Storage of Liquids, Cement and Concrete Association of New Zealand, Standards Association of New Zealand, 2009.
173. Recommended Standards for Water Works, Great Lakes- Upper Mississippi River Board of State and Provincial Public Health and Environmental Managers, Health Research Inc., Health Education Services Division, POBox 7126, Albany, NY 12224, 2012 102-SA Water TG.
174. Planning and Infrastructure, Above Ground Circular Concrete Tank, Manager Engineering, 2008. 103- Selected Earthquake Engineering Papers of George W. Housner, American Society of Civil Engineers, 1990.

175. Specification for Construction of Concrete Reservoirs, Standard Specifications and Drawings Specification for Construction of Concrete Reservoirs, Gold Coast, Australia, SS 11, 2004.
176. TM 5-809-10, Seismic Design Guidelines for Essential Buildings, US Department of the ARMY, Washington, D.C., 1986.
177. USEPA Finished Water Storage Facilities, U.S. Environmental Protection Agency, 2002.
178. Zecchin, A. C.; Maier, H. R.; Simpson, A. R.; Leonard, M.; Nixon, J. B. Parametric Study for an Ant Algorithm Applied to Water Distribution System Optimization. *IEEE Trans. Evolut. Comput.* **2005,** *9* (2), 175–191.
179. Walski, T. M. Hydraulic Design of Water Distribution Storage Tanks. In *Water Distribution Systems Handbook*, Chapter 10; Mays, L. W., Ed.; McGraw-Hill: New York, 2000,
180. Farmani, R.; Walters, G.; Savic, D. Evolutionary Multi-objective Optimization of the Design and Operation of Water Distribution Network: Total Cost vs. Reliability vs. Water Quality. *J. Hydroinform.***2006,** *8* (3), 165–179.
181. Farmani, R.; Savic, D. A.; Walters, G. A. The Simultaneous Multi-Objective Optimization of Anytown Pipe Rehabilitation, Tank Sizing, Tank Siting and Pump Operation Schedules; Critical Transitions in Water and Environmental Resources Management, ASCE, American Society of Civil Engineers: Salt Lake City, 2004; pp 4663–4672.
182. Haarhoff, J.; Van Zyl, J. E.; Nel, D.; Van der Walt, J. J. Water Supply and Water Quality in Rural Areas: Stochastic Analysis of Rural Water Supply Systems. *Water Supply* **2000,** *18* (1), 467–470.
183. Maier, H. R.; Simpson, A. R.; Zecchin, A. C.; Foong, W. K.; Phang, K. Y.; Seah, H. Y.; Tan, C. L. In *Ant Colony Optimization for the Design of Water Distribution Systems*, Proceedings of World Water and Environmental Resources Congress 2001; ASCE, The Rosen Plaza Hotel: Orlando, Florida, United States, 2001.
184. Maier, H. R.; Simpson, A. R.; Zecchin, A. C.; Foong, W. K.; Phang, K. Y.; Seah, H. Y.; Tan, C. L. Ant Colony Optimization for Design of Water Distribution Systems. *J. Water Resour. Plan. Manag.* **2003,** *129* (3), 200–209.
185. Walski, T. M.; Chase, D. V.; Savic, D. A.; Grayman, W.; Beckwith, S.; Koelle, E. In *Advanced Water Distribution Modeling and Management*; Haestad Press: Watertown, Conn, 2003.
186. Hashemi, S. S. Optimization of Water Networks with Minimization of Pumping Energy. Master Thesis, University of Tehran, Iran, 2010.
187. Mau, R. E.; Boulos, P. F.; Clark, R. M.; Grayman, W. M.; Tekippe, R. J.; Trusell, R. R. Explicit Mathematical Models of Distribution Storage Water Quality. *ASCE J. Hydraulic Eng.* **1995,** *121* (10), 669–709.
188. Rossmann, L. A. EPANET 2 User´s Manual. Water Supply and Water Resources Divison; Environmental Protection Agency: Cincinnati, EE.UU, U.S., 2000. http://www.epa.gov/ORD/NRMRL/wswrd/epanet.html
189. Rossmann, L. A.; Grayman, W. M. Scale-Model Studies of Mixing in Drinking Water Storage Tanks. *J. Environ. Eng.* **2000,** *125* (8), 755–761.
190. Yeung, H. Modeling of Service Reservoirs. *J. Hydroinform.* **2001,** *3* (3), 165–172.
191. Martínez-Solano, F. J.; Lopez-Jimenez. P. A; Iglesias-Rey, P.; Perez, R. 2006. "Utilization of fluid dynamics co
192. Mahmood, F.; Pimblett, J. G.; Grace, N. O.; Grayman, W. M. Evaluation of Water Mixing Characteristics in Distribution System Storage Tanks. *J. AWWA* **2005,** *97* (3), 74–88.

193. Marek, M.; Stoesser, T.; Roberts, P. J. W.; Weitbrecht, V.; Jirka, G. H. In *CFD Modeling of Turbulent Jet Mixing in a Water Storage Tank*, XXXII IAHR Congress, Venice (Italy), July 1/6, 2007, 2007.
194. Van Zyl, J. E.; Haarhoff, J. Reliability Analysis of Municipal Storage Reservoirs Using Stochastic Analysis. *J. S. African Inst. Civil Eng.* **2007,** *49* (3), 27–32.
195. Leon, A. S. Improved Modeling of Unsteady Free Surface, Pressurized and Mixed Flows in Storm-Sewer Systems. Submitted in Partial Fulfillment of the Requirements for the Degree of Doctor of Philosophy in Civil Engineering in the Graduate College of the University of Illinois at Urbana-Champaign, 2007; pp 57–58.
196. Hariri, A. K.; Nagiyev, F. B.; Haghi, A. K. Physical Modeling of Fluid Movement in Pipelines. In *Nanomaterials Yearbook*, 2009. USA.https://www.novapublishers.com/catalog/product_info.php?products_id=11587
197. Leon, A. S.; Ghidaoui, M. S.; Schmidt, A.; Garcia, M. H.; Godunov Type Solutions for Transient Flows in Sewers. *J. Hydraulic Eng.* **2006,** *132* (8).

CHAPTER 6

Bioapatite from Aquatic Resources as Remedial Matrix for Water Purification

P. K. BINSI and S. J. LALY

ICAR—Central Institute of Fisheries Technology, Matsyapuri, Willington Island, Cochin, India

ABSTRACT

Hydroxyapatite ($Ca_{10}(PO_4)_6(OH)_2$ (HAP) has gained widespread attention as a green and environmentally friendly adsorbent due to its excellent ability in removing several organic and inorganic pollutants from waste water and natural water bodies. The unique hexagonal crystal structure of HAP along with abundant reactive moieties supports the quenching of pollutants from aqueous media particularly through an adsorption mechanism. Both chemical and physical forces play a major role in the removal of pollutants from water media, including ionic exchange reaction, surface complexation, coprecipitation, electrostatic interaction, and hydrogen bonding. The efficacy of pollutant removal may be enhanced by surface modification and composite formation with synergistic adsorbents. Magnetic HAP is such an option as it also provides a solution for the regeneration and recovery of HAP from treatment system and hence may be reused in subsequent cycles. The first part of this chapter is devoted to natural hydroxyapatite and extraction of biological hydroxyapatite from aquatic sources. The second part of this chapter focuses on its use as remedial material for water purification.

Sustainable Water Engineering: Smart and Emerging Technologies. Ali Pourhashemi, Kian Hariri Asli, Ann Rose Abraham, & A. K. Haghi (Eds.)

6.1 INTRODUCTION

Water bodies are alarmingly polluted by intensive agricultural and industrial production activities all over the world. Pollutants include inorganic metal ions, nutrients, synthetic organic compounds such as polyaromatic hydrocarbons, dyes, pesticides, insecticides, petroleum by-products, pharmaceuticals, endocrine disruptors etc. (Gupta et al., 2015). Improper disposal of electronic waste is also a major activity contributing to the increased level of heavy metal contamination, and these inorganic pollutants are highly stable compared to organic contaminants like pesticides and products derived from petroleum. The toxicity levels of these pollutants vary with the nature, type, concentration, and degradation pattern of the pollutants. Exposure to these pollutants via drinking water results in incidence of different diseases spanning from mild skin allergies to severe organ disorders and even death in living beings (Sánchez-Bayo, 2011). Adsorbents having favorable surface physico-chemical features aid the removal of different pollutants from water. Enhanced performance of adsorbent materials is necessary to meet the complex nature of pollutants in water.

Hydroxyapatite ($Ca_{10}(PO4)_6(OH)_2$) (HAP), a major member of apatite family, is a promising candidate for eco-remediation owing to its excellent adsorption capacity, acid–base adaptability, ion exchange capacity, and thermal stability. The conventional routes for the synthesis of hydroxyapatite use toxic solvents and toxic additives to achieve expected crystallinity and morphology. On the other hand, hydroxyapatite of biological origin follows greener route of extraction and yields apatite with adequate properties.

6.2 HYDROXYAPATITE

The name "apatite" refers to a family of compounds with a common structure, that is, hexagonal system, space group, P63/m. Apatites can be of natural/biological origin or can be synthesized in the laboratory. Natural hydroxyapatite can be of biological origin (mammalian bones, fish scale, fish bone, exoskeletons of crustaceans, molluscs, corals, egg shell, plants and algae etc.) or from mineral reserves (e.g., limestone). The synthetic HAP having desirable Ca/P value can be prepared in the laboratory from calcium and phosphate compounds via chemical precipitation method in solutions containing different kinds of elements such as CO_3, F, Mg, Sr, and Cl.

The quantity and kind of substitution in the apatite are determined by the preparation process. Among the various synthesis protocols, hydrothermal-assisted extraction process, chemical precipitation, and extraction from biogenic sources are the most widely cited protocols in recent years. In synthetic apatites, calcium phosphates change from one type to another and then to apatite through a process of dissolution and precipitation, but in biological apatites, the amount of dissolution in acid buffers seems to depend on the composition of the apatites (Strunz and Nickel, 2001). HAP extracted from biowaste, known as bioapatites, can be extracted following a nontoxic and greener pathway of synthesis and are renewable with higher availability at an affordable cost.

6.3 BIOLOGICAL HYDROXYAPATITE

In recent years, greater emphasis has been given to adopt green production approaches due to rising concerns about carbon foot printing and global warming. Various sources are found to be suitable for the extraction of natural HAP, such as terrestrial and marine resources. Several studies were conducted on the extraction of biological HAP from bioresources such as coral, bovine bone, fish bone, fish scale, and molluscan shells (Karacan et al., 2021; Bano et al., 2017; Binsi et al: 2022; Alif et al., 2018). Among the mammal sources, most of the studies used bovine bone as the raw material. Prior to HAP extraction, boiling and washing with alkali and solvents is desirable for the removal of fat and soft tissues from animal bones (Bano et al., 2017). Vegetative parts of plants and algae are also considered potential sources of HAP. Vegetable and fruit processing discards such as leaves, stalk, flower, fruit peel, and vegetable peel have been extensively investigated for HAp extraction (Robinson et al., 2018). Marine species like corals and red algae (Phymatolithon Calcareum) contain calcite (CaCO3). Red algae calcite is converted into hydroxyapatite via pyrolysis and hydrothermal method.

The majority of biological apatites are associated with a number of minor and trace elements and are calcium-deficient and non-stoichiometric. The formation of tricalcium phosphate (TCP) crystals along with HAP following sintering also reflects the non-stoichiometry of these apatites. The absence of specific trace elements like Na, Mg, and Al, which are frequently associated with natural apatite, may prevent synthetic HAP from accurately simulating the characteristics of natural apatite, though. Since carbonate is present in the lattice structure of biological apatites, biological apatites are usually categorized as carbonate apatites. The CO_3 content in apatite is significantly lower when fluoride is present, as observed in some calcified fish scales and molluscan shells. Magnesium is another minor component of biological apatites. The F content of hard tissues of shark and certain fish species ranged from 0.3 to 3.9% (wt%), which was unrelated to the F content in the inhabited water (Lee and LeGeros, 1985). However, an inverse relationship between F and CO3 content could be observed. The F content in biological apatite was associated with increased crystallinity and decreased dissolution of apatite in acid buffer (Legeros, 1981). Basically, HAP comprises a lattice of hydroxide (OH–) ions aligned in columns parallel to the c-axis, along with their Ca2+ and PO4 3 counterparts. Biological apatites show significant variations in lattice dimensions as well; Shark enamel apatite has a smaller a-axis (0.9381 nm) than pure HAp, whereas human enamel apatite has a substantially bigger a-axis (0.944 nm compared to 0.942 nm) (LeGeros and Suga, 1980). The extraction of hydroxyapatite from biowastes offers additional advantage in that it contributes to value addition of biowaste. Process discards rich in hydroxyapatite and calcium are animal bones and teeth, fish scale and bone, cuttlefish bone, molluscan shell etc. Out of the different bioresources, aquatic resource is widely preferred for various reasons, including the abundance and cost-effectiveness.

6.4 EXTRACTION OF BIOLOGICAL HAP FROM AQUATIC RESOURCES

Several methodologies are documented in the literature for the extraction of HAP from various biowastes such as sintering or calcination, hydrolysis, microwave, ultrasound, and combination of these (Figure 6.2). A description of the common protocols documented in the literature is given in Table 6.1.

The natural sources of HAP can be either direct sources of HAP or sources of Ca/P. For instance, animal bones, fish scale, etc. are direct sources of biological HAP, whereas egg shell, molluscan shell, crustacean shell, etc. which are indirect sources of HAP contain calcium as high as 95% and above as CaCO3. Similarly, there are natural mineral reservoirs, which provide calcium and phosphate for HAP synthesis (Figure 6.1). In the case of molluscan shell, initially $CaCO_3$ is converted into CaO by calcination at elevated temperature and thereafter converted into HAP by reacting with phosphoric acid. Alternatively, other sources of phosphorous such as di-ammonium phosphate under hydrothermal condition can be used for the synthesis of HAP from corals and molluscan shells.

TABLE 6.1 Extraction Protocols Documented in the Literature for Hydroxyapatite from Aquatic Animals.

Source	Extraction method	Application	References
Tuna bone Sword fish bone	Thermal calcination	Biomedical	Boutinguiza et al. (2012)
Salmon bone	Alkaline hydrolysis	Biomedical	Venkatesan et al. (2015)
Tilapia bone and scale	Thermal calcination at 800–1000°C	Biomedical	Mustafa et al. (2015)
Fish bone	Thermal calcination	Biomedical	Sunil and Jagannatham (2016)
Lates calcarifer fish bone	Thermal calcination at 1200°C	Biomedical	Pal et al. (2017)
Fish bone	Thermal calcination	Nickel adsorption	Daibiri et al. (2018)
Rainbow trout bone	Thermal calcination	Biomedical	Shi et al. (2018)
Skipjack Tuna bone	Precipitation	Tissue engineering	Wardani et al. (2020)
Fish bone	Thermal calcination	Heavy metal adsorption	Hernández-Cocoletzi et al. (2020)
Snakehead fish bone	Co-precipitation	Phenol adsorption	Hariani et al. (2020)
Tuna fish bone	Thermal calcination	Cadmium adsorption	Herawati et al. 2020
Fish bone (*Sardinella longiceps*)	Alkaline hydrolysis	biomedical	Surya et al. (2021)
Fish bone	Thermal calcination	Cadmium removal	Foroutan et al. (2021)
Lates calcarifer bone	Thermal calcination	Biomedical	Le Ho et al. (2022)
Tilapia fish scale	Alkaline hydrolysis	Chromium removal	Selimine et al. (2022)
Fish scale	Alkaline hydrolysis	Bisphenol A biosorption	Aziz et al. 2022
Tilapia scale	Thermal calcination	Water treatment	Liaw et al. (2020)
Crucian carp scale	Deep eutectic solvent extraction	Metal ion adsorption	Liu et al. (2021)
Fish scale	Ionic liquid pre-treatment	Biomedical	Muhammad et al. (2016)
Rohu scale	Thermal calcination	Heavy metal removal	Fijol et al (2022)
Fish scale	Ultrasound irradiation	Dye removal	Sricharoen et al.(2021)

TABLE 6.1 *(Continued)*

Source	Extraction method	Application	References
Nile Tilapia fish scale	Sonochemical coprecipitation	Hg^{2+} removal	Srichroen et al. (2020)
Fish scale	Alkaline heat treatment	Bio-supercapacitor	Eswaran et al., 2021
Fish scale	Microwave irradiation	Ammonia removal from aquaculture pond	Hamzah et al. (2019)
Fish scale	Calcination	Azo dye removal	Das et al. (2019)
Marine Coral	Hydrothermal process	Biomedical	Nandi et al.(2015)
Hydrozoan skeleton	Hydrothermal process	Drug delivery	Karakan et al. (2021)
Cuttlefish bone	Precipitation	Biomedical	Venkatesan et al.(2018)
Cuttlefish bone	Coprecipitation	Biomedical	Faksawat et al. (2015)

FIGURE 6.1 Major aquatic sources of HAP.

Most of the investigations with direct sources of HAP employ thermal calcination as the primary extraction method. Several documented studies suggest higher calcination favors improving the mechanical strength of resultant HAP (Choi et al., 1999). The high temperature is also required for the elimination of collagen, so as to release hydroxyapatite crystals in pure form. Sintering at a temperature above 700°C was reported as optimal for obtaining microporous and crystalline HAp (Krishnamurithy et al. 2014), 900–1200°C for cod bone (Piccirillo et al., 2013). Pal et al. (2017) demonstrated the formation of HAP crystals of various dimensions in the temperature range of 200–1200°C, wherein the powders sintered above 800°C presented irregular morphology and composition. Ozawa and Suzuki (2002) suggested a sintering temperature close to 1300°C for fish bone as an ideal temperature for obtaining crystalline HAP with required porosity.

A pretreatment with NaOH and H_2O_2 followed by sintering at 400–1000°C was suggested for mango fish bone (*Oreochromis niloticus*) (Bunsiri et al., 2012). The authors have reported a concurrent observation that organic phase could be completely eliminated by annealing at 600°C and above, wherein the crystallinity of HAP increased gradually in the temperature range of 400 to 600°C, significantly from 600 to 800°C and showed a plateau phase from 800 to 1000°C. Venkatesan and Kim (2010) suggested an isolation temperature of 600–900°C as the most ideal condition for the extraction of

HAP from tuna (*Thunnus obesus*) bone. Likewise, Prabakaran and Rajeswari (2006) reported 900°C as the optimum temperature for the formation of phase pure fish bone HAP. A calcination temperature of 850°C for 4 h was opted for the extraction of HAP from Atlantic Bonito (*Sarda sarda*) bone, which contained 66.7% HAP and 33.3% tricalcium phosphate (TCP) (Gunduz et al., 2017). Boutinguiz et al. (2012) extracted B type hydroxyapatite from bones of sword fish and tuna by heat incineration at 600°C, whereas at 950°C a biphasic material with biological hydroxyapatite/beta-TCP at a ratio of 87/13 was obtained. A calcination temperature of 900°C for 8 h was suggested for tilapia bone by Da Cruz et al (2010). A more recent study in our laboratory indicated complete desorption of residual collagen from carp scale surface at 800°C, and that from matrix interspace at 900°C (Binsi et al., 2022). The SEM, AFM, and electrochemical impedance spectroscopy confirmed stable HAp nanocrystal formation at 1000°C, above which the characteristic features of recrystallization and phase transformation were evident. Hence, 1000°C was suggested as the rate-limiting temperature for HAP formation (Figure 6.2).

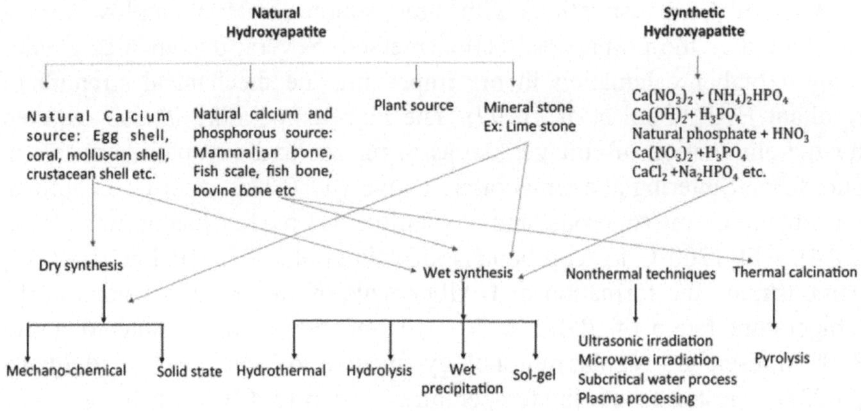

FIGURE 6.2 Common protocols reported in the literature for the synthesis of hydroxyapatite.

Calcination under severe conditions poses environmental challenges; hence, alternative strategies were adopted by several researchers for the removal of organic phase, collagen. Huang and Chu (2013) suggested enzymatic hydrolysis prior to heat sintering to obtain porous HAP from fish scale. High-frequency microwave (MW) sintering at 30 GHz was attempted by Hassan et al. (2016), and reported that MW irradiated HAP exhibited higher porosity and mechanical properties, compared to those of heat-sintered

HAP. Hamzah et al. (2019) demonstrated higher adsorption capacity toward ammonia from aquaculture waste water for HAp extracted from fish scale via microwave irradiation. Sonochemical coprecipitation methodology was employed to extract. HAp from Nile tilapia was bone by Sricharoen et al. (2020). Athinarayanan et al. (2020) proposed a hydrothermal process for the simultaneous extraction of carbon dots (CDs) and rod shaped and spherical HAP particles of nanodimensions. Later, Ashraf et al. (2021) suggested a process to cosynthesize nano carbon dots and nano β-dicalcium diphosphate from the scale of Labeo rohita along with hydroxyapatite. In the process, collagen in fish scale is converted into nano carbon dots by a hydrothermal process using acetic acid whereas the mineral residue was converted into hydroxyapatite. The hydroxyapatite thus formed was found to have superior crystalline and morphological properties compared to that formed by the conventional sintering process. A green protocol was proposed by Liu et al. (2021) to extract hydroxyapatite from crucian carp scales using eutectic solvent mixture of choline chloride and 1,4-butanediol. The extracted HAp exhibited superior affinity toward metal ions such as lead, copper, silver, and zinc. In brief, the available reports on aquatic HAp suggest that the calcination regime is to be carefully designed for the efficient removal of the organic matter from the native form, as well as for obtaining HAp particles with desired crystallinity and porosity, while avoiding thermal decomposition of the final product.

6.5 HYDROXYAPATITE FOR WATER PURIFICATION

HAp has become more popular among solid sorbents due to its excellent absorption capacity, ionic exchange characteristic, low water solubility, and proven efficiency for the removal of heavy metal ions, fluorides, and dyes from aqueous solutions (Table 6.2). Apart from that, nontoxicity, the excellent chemical stability under reducing and oxidising conditions, high thermal stability, and water insolubility are added advantages of HAP (Ibrahim et al., 2020).

The superior properties of hydroxyapatite as a pollutant remediation matrix arise from its chemical nature, with both acidic and basic sites, which enables HAP to act as a heterogeneous catalyst. The catalytic activity of hydroxyapatite is mainly related to its Ca/P ratio. The acid site density may be increased by increasing the Ca/P ratio, and vice versa. When the Ca/P ratio is around 1.5, hydroxyapatite exhibits acidic nature and at 1.67, basic behavior is exhibited. Both acidic and basic attributes are exhibited when the ratio is between 1.50 and 1.67.

TABLE 6.2 Selected Literature on Pollutant Removal by HAP.

Material	Pollutant	Adsorption capacity (mg/g)	References
Heavy metals			
HAP scaffold	Lead	208.3 (pH = 5) 344.8 (pH = 2.2)	Liu et al. (2017)
Magnetic HAP	Cadmium	84.8	Kazeminezhad et al. (2014)
HAP	Lead, Cadmium,	Pb^{2+}: 700	Mobasherpour et al. (2012b)
	Nickel	Cd^{2+}:142	
		Ni^{2+}:36.25	
HAP nanofibers	Copper, Cadmium, Lead	Cu^{2+}: 23	Zhou et al. (2019)
		Cd^{2+}: 36	
		Pb^{2+}: 93	
Chitosan/nanohydroxyapatite composite	Mercury	127.4	Hassan and Hrdina (2018)
n-HAP/chitosan composite	Iron	6.753	Kousalya et al. (2010)
Chitosan/Ag-hydroxyapatite nanocomposite	Copper	40.11	Li et al. (2018)
Magnetic nano-hydroxyapatite encapsulated alginate beads	Chromium	29.1	Periyasamy et al. (2018)
Hydroxyapatite/calcium silicate hydrate composite adsorbent	Lead	946.7	Zhang et al. (2018)
Hydroxyapatite nanoparticles	Arsenic	526	Mirhosseini et al. (2014)
Dyes			
HAP	Methylene blue	14.27	Wei et al. (2015)
HAP solutions	Solophenyl Turquoise	1687	Lemlikchi et al. (2012)
Chitosan-doped HAP	Congo red	769	Hou et al. (2012)
Chitosan coated magnetic hydroxyapatite	Reactive blue 19	26.6	Nguyen et al. (2014)

TABLE 6.2 *(Continued)*

Material	Pollutant	Adsorption capacity (mg/g)	References
Chitosan/Ag-hydroxyapatite nanocomposite	Rhodamine B	127.61	Li et al. (2018)
Magnetic HAP	Methylene blue	328.4	Wang et al. (2017)
Sodium alginate HAP	Acid Yellow 220	212.8	Manatunga et al. (2018)
Polyalcohol-coated HAP	Congo red, Methyl blue,	170.7	Guan et al. (2018)
		379.1	
	Methyl orange	14.7	
Fluoride ions			
Hydroxyapatite (HA) multiwalled carbon nanotubes (MWCNTs)	Fluoride	30.22	Ruan et al. (2017)
Non-calcined synthetic hydroxyapatite	Fluoride	4.38	Singh et al. (2020)
Magnetic nano-hydroxyapatite/alginate composite	Fluoride	4.05	Pandi and Viswanathan (2015)
Hydroxyapatite-modified activated alumina	Fluoride	14.4	Tomar et al. (2015)
Radionuclides			
Nanocrystalline bio-hydroxyapatite	Sr (II)	5.35	Handley-Sidhu et al. (2011)
In situ-grown nanohydroxyapatite on magnetic Ca Al-layered double hydroxides	U (VI)	261.1	Krestou et al. (2004)
High-surface-area hydroxyapatite by in-situ precipitation	Sr (II)	28.51	Kim et al. (2019)
Hydroxyapatite beads	Co (II)	7.73–11.35	Tran et al. (2021)
Phenol			
Hydroxyapatite-loaded films based on cellulose nanofibrils	Phenol	30–64	Narwade et al. (2017)

TABLE 6.2 *(Continued)*

Material	Pollutant	Adsorption capacity (mg/g)	References
Pesticides			
Magnetite hydroxyapatite nanocomposite	Paraquat dichloride	416.67	Kotnala et al. (2022)
Nano-hydroxyapatite	Atrazine	Photo degradation process	Rubi et al. (2020)
Nano hydroxyapatite modified CFGO/ZnO nanorod composite	Chlorpyrifos	Photo degradation process	Anirudhan et al. (2021)
Drugs			
Hydroxyapatite@Montmorillonite hybrid composite	Ciprofloxacin	91.18	Laabd et al. (2021)
Synthetized materials from Hydroxyapatite	Oxytetracycline	57–61	Martínez-Olivas et al. (2020)
Hydroxyapatite by Fe(III) incorporation	Tetracycline	45.39	Li et al. (2017)
Hydroxyapatite (HA) nanopowders	Oxytetracycline	291.32–278.27	Harja and Ciobanu (2018)
Zeolite-hydroxyapatite-activated oil palm ash	Tetracycline	186.09–244.63	Khanday and Hameed (2018)
Ultra-high molecular weight polyethylene based nano-composite with nano-hydroxyapatite and organophilic bentonite	Tetracycline	8.72	Fagundes et al. (2019)
Nitrobenzene			
Nanocrystalline hydroxyapatite	Nitrobenzene	8.993	Wei et al. (2010)
Hydroxyapatite–gelatin nanocomposite	Nitrobenzene	42.373	Wei et al. (2014)

The acidic behavior of HAP is related to HPO42- group (Bronsted acid site) and Ca2+ and OH- vacancies) (Lewis acid site) (Silvester et al., 2014). Hence, calcium-deficient HAP is considered to have more acid sites. The ion-exchange capability of HAP mainly arises from its amphoteric nature, having both cationic and anionic sites. The Ca2+ ion can be substituted by alkaline earth metals (Mg2+, Sr2+, etc.), transition metal ions (Co2+, Cu2+, Zn2+, Mn2+, etc.), and other cations like A13+, Pb2+, etc. (Wakamura et al., 2000). The charge density off cations largely determines the ion exchange capacity of HAP with Ca2+ ion. The anionic exchange involves the substitution of OH- by anions such as F-, Br-, CO32-, etc. or PO4 2- by HPO4 2-,SO4 2-, CO3 2-, etc. (Lafon et al., 2008). The physicochemical properties of HAP can thus be altered through the ion exchange mechanism, which can be further utilized for enhancing the catalytic performance of HAP (Pai et al., 2020).

6.5.1 MECHANISM OF ACTION

Techniques like adsorption, membrane technology, ion exchange, coagulation, electrochemical, or oxidation are applied for contaminant removal from the environment. Each one of these treatments has merits and demerits. Adsorption technique is highly accepted for the treatment of organic and inorganic contaminants in water. Adsorption has advantage of reduced cost, increased efficiency, and greater flexibility (Chen et al., 2016). Materials like chitosan, bark, dead biomass, bone char, natural zeolites, clay, agricultural wastes, waste materials from industrial origin like fly ash and oxides are categorized as low-cost adsorbents.

Application of different inorganic materials is successfully used to remove harmful contaminants from wastewater originated from agricultural, municipal, industrial, and nuclear sources (Rajec et al., 1996). Different adsorbents such as activated carbon, metallic oxides and oxyhydroxides, carbonates, and calcined phosphates are used for different water treatment applications. Among these calcined phosphates have shown a promisable and important role in environmental remediation because of chemical, surface, and structural properties (Anandan and Janakiram, 2012).

HAP has a hexagonal crystalline structure containing calcium and phosphorous at a ratio of 1.67. It has high bioactivity (Maghsoodi et al., 2020) and has different mechanisms for pollutant removal. The common mechanisms are dissolution precipitation involving the precipitation of metal ions, ion exchange through substitution of ions, physical adsorption using different adsorbents, and electrostatic interactions for adsorbing ions (Figure 6.3).

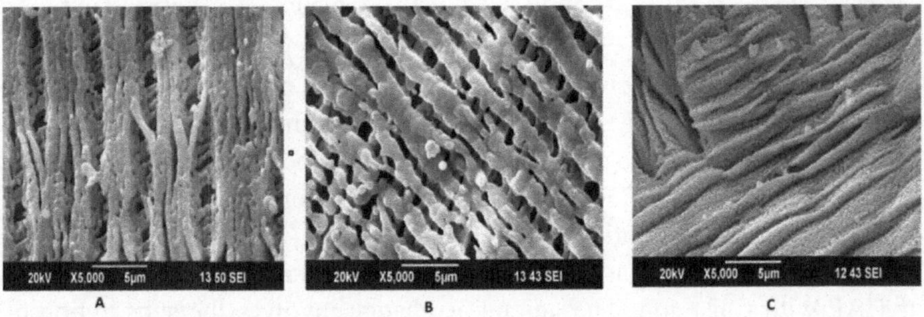

FIGURE 6.3 Scanning electron microscopic images of HAP extracted in various sintering conditions: (A) 900°C; (B) 1000°C; and (C) 1100°C.

6.5.1.1 ADSORPTION

In comparison to conventional chemical precipitation, advanced oxidation, electrochemical and membrane technologies, adsorption is a clean methodology which can be applied in water treatment economically (Gupt et al., 2015). Inorganic adsorbents like zeolites, metal oxides, clays etc. are apatites with distinct crystal structure and efficient adsorption potential (Mehta et al., 2018). The HAP has excellent surface properties which can be effectively applied to reduce the level of harmful pollutants in water. The multifunctional role of HAP in environmental remediation was reported previously (Ibrahim et al., 2019; Pai et al., 2020). Presence of heavy metals is a serious issue affecting all living beings because of non-biodegradability and toxicity at lower concentration. In adsorption, heavy metal ions accumulate as a monolayer on the adsorbent and can be removed easily. Adsorption is an easy technique and efficient for removing metal ions from water especially using natural polymeric materials. Also, huge quantity of polluted water can be treated in a short span of time. HAP is a suitable sorbent for the control of pollutants due to unique properties such as absence of harmful effects, low cost involved and ease of availability, suitability in adsorption, and increased stability during different chemical environments (Ibrahim et al., 2019). Generally, larger surface areas (minimum particle size) favor the process of adsorption as it favors multilayer adsorption through surface-active forces. The weak nature of these forces makes the process of removal of contaminants easy (Deng et al., 2019) and this helps in the recovery of HAP and its reuse for further adsorption processes.

The adsorption process in the case of HAP works primarily through the surface complexation mechanism through phosphate, calcium, and hydroxyl groups. The positively charged calcium ions provide anion adsorption sites and the negatively charged groups such as carboxylic acid or phosphate act as cation adsorption sites. Pollutants such as heavy metals, free radical ions, organic pollutants, etc. are removed via this surface complexation process. In addition to complexation, physical adsorption through electrostatic interaction also plays a major role in adsorption. The exposed OH^- groups in HAP lattice form hydrogen bonds with certain specific pollutants such as certain drugs, dyes, and phenol.

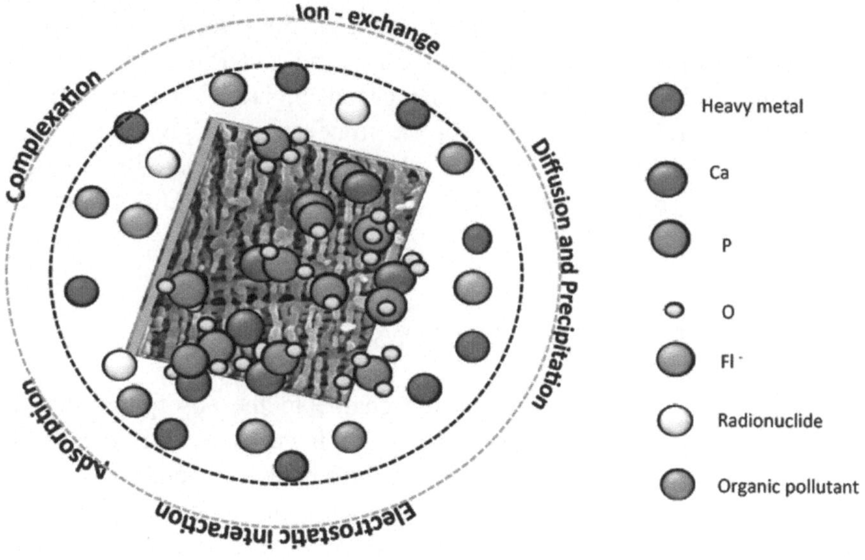

FIGURE 6.4 Possible mechanism of pollutant removal by HAP, as per available literature.

6.5.1.2 DISSOLUTION PRECIPITATION

Dissolution of HAP results in the saturation of phosphate ions which can precipitate metal cations. Phosphate ions bind with metal ions resulting in the formation of stable metal crystals. The mechanism involved is immobilization of metal ion by replacing the metal component in HAP using Ca^{2+} ion, for instance, the removal of lead ions by HAP is achieved by complexing with lead ions to form lead–phosphate complex. Anions can influence hydroxyl metal crystal phase without making changes in the metal adsorption by HAP (Ibrahim et al., 2020) and the nature of the crystal formation varies with the type of anion present. HAP can immobilize Pb^{2+} cations in the presence of

NO^3, Cl, F, $SO_4{}^2$, and $CO_3{}^2$ anions. Vila et al. (2011) reported higher removal efficiency for Pb^{2+} at low pH through the process of dissolution precipitation and a lower removal efficiency at higher pH.

6.5.1.3 ION EXCHANGE

HAP lattice can undergo replacement with either cations (Ca^{2+}) or anions ($PO_4{}^{3-}$ and/or OH). Calcium ions will be replaced by transition metals, alkaline earth, and other metal types through ion exchange. Removal and recovery of metals via ion exchange is one of the effective methods for reducing the pollution (Ibrahim, 2009). Ca (I) and Ca (II) in the HAP lattice will be replaced by cations in which larger ions replace Ca (II) and smaller ions replace Ca(I) with respect to the electronegativity (Opre et al., 2005). Metal replacement in HAP's structure is based on electronegativity as well as ionic radius. The replacement of Cd^{2+}, Sr^{2+}, and Pb^{2+} in the HAP is primarily determined by Ca (II) sites.

6.5.1.4 ELECTROSTATIC INTERACTIONS

Substances which are having opposite charge to the adsorption sites in HAP can be adsorbed through electrostatic interactions. OH^- present in the exposed crystal surfaces of HAP can form hydrogen bonds and can be applied for the removal of halogen ions and pollutants of organic nature. The HAP and the pollutant interact via electrostatic forces during the adsorption process (Yin et al., 2002). Van der Waals bonds in electrostatic interactions help in adsorbate ion removal. pH determines the charge of HAP and when pH is less than pH_{PZC} at which the net charge is zero, it forms a positively charged surface. Generally, adsorption is increased with the decrease in pH (Sundaram et al., 2008).

6.5.2 APPLICATIONS OF HAP TO REDUCE ENVIRONMENTAL POLLUTANTS

HAP is an environment-friendly material for pollutant removal as it is nontoxic and biocompatible. Eco-friendly and low-cost multifunctional micro and nanocomposite materials for the removal of heavy metals, dyes, and microbes from solutions are reported. The HAP contain 2.6 P-OH groups nm^{-2} which forms the binding sites for CO_2, CH_3OH, H_2O, $CH_3(CH_2)_3NH_2$, and CH_3COOH. Efficient surface properties make HAP suitable for elimination of environmental pollutants. A brief account of HAP structures documented in the literature for water treatment is presented in Table 6.2.

6.5.2.1 REMOVAL OF HEAVY METALS

Toxic effect of heavy metals at very low concentrations is also a threat of concern to the aquatic environment and associated organisms. Industrial discharge of waste waters containing heavy metals is alarmingly troublesome in drinking water, agriculture irrigation, soil etc. HAP can be considered an advanced adsorbent for the elimination of heavy metals and HAP composites are more efficient in adsorption compared to HAP alone. The metal trapping mechanism is largely dependent on metal species, the treatment conditions like pH and contact time, and the surface characteristics (acid–base sites ratio) of HAP (Campisi et al., 2018). Ca(I) or Ca(II) ions of HAP can be completely or partially replaced by metal ions (Ferri et al., 2019). There are several reports on the efficiency of HAP in immobilizing ions of chromium, lead, cadmium, nickel, zinc, aluminum, copper, iron, cobalt, manganese etc. (Ferri et al., 2019; Selimin et al., 2022). The adsorption of chromium using magnetic HAP is well studied. Doping of HAP with cobalt supports the adsorption of cesium with a maximum of 69.5 mg/g in which the doping improved the adsorption efficiency compared to nondoped HAP. It also has superior adsorption property toward some cations compared to others due to greater affinity (Mobasherpour et al., 2012a). The binding of bivalent metal cations has a positive correlation to its characteristics and depends on groups such as hydroxyl, phosphate, and carbonate. Calcium ions get substituted with metal cations like lead, zinc, cadmium, copper, cobalt, and nickel; phosphate groups can be replaced with anions like AsO_4^{3-}, VO_4^{3-}, NO^{3-}, PO_4^{3-}, CO_3^{2-}, SO_4^{2-} etc, while hydroxyl groups substitute anions (F^-, Cl^-) (Lin et al., 2009).

Removal of arsenate from contaminated drinking water to the extent of 88% could be achieved using HAP nanoparticles at a level of 0.6 g/L (Mirhosseini et al. 2014) at higher pH, while multi-layer adsorption showed a higher capacity of adsorption at 526 mg/g. Nanocrystalline hydroxyapatite exhibited higher efficiency compared to activated carbon in eliminating Pb^{2+} and Zn^{2+} (Stötzel et al., 2009). Removal efficiency of 97.3 to 98.6% for copper and 94.2 to 97.7% for zinc was reported by Stötzel et al. (2006). In multi-metal systems, competition among metals reduces the metal removal efficiency of HAP. Lead in solution can be easily removed by dissolution precipitation and ion exchange mechanisms. Most of the heavy metal adsorption processes by HAP follow pseudo-second-order kinetic model implying the control of sorption process by different mechanisms with a lesser extent of physical adsorption (Brazdis et al., 2021). HAP follows the Langmuir isotherm in the case of monolayer adsorption and increase of temperature can increase the adsorption process. At reduced pH, the solubility of HAP increases due to increased hydrogen ion concentration. It is also associated with the decrease of OH^- and PO_4^{3-} ions. According to Campisi et al. (2018), the Ca/P molar ratio of HAP is also crucial in deciding the metal removal efficiency.

Combination with magnetic materials facilitates the separation of the adsorbent and illustrates good performance in regard to the removal of the pollutants. Magnetic HAP nanoparticles have a higher efficiency (96%) in removing cadmium ions with an adsorption capacity of 84.8 mg/g, while HAP nanofibers showed an adsorption capacity of 76.9 mg/g (Pai et al., 2020). Magnetic nano HAP encapsulated with alginate has shown chromium adsorption capacity of 29.1 g/g (Periyasamy et al., 2018). HAP and calcium silicate hydrate composite is efficient in removing lead ions at an adsorption capacity of 946.7 mg/g (Zhang et al., 2018). The coating of HAP nanoflakes and turmeric extract on granular activated carbon efficiently removed Pb^{2+} at room temperature at different time intervals and different pH levels (Jayaweera et al., 2018). Magnetic properties can prevent the loss of adsorbent to the environment and further chance of contamination.

6.5.2.1.1 Modeling of Heavy Metal Adsorption

Adsorption processes can be evaluated based on kinetic models, equilibrium isotherms, and thermodynamic data. Adsorption processes can be optimized in correlation with operational parameters of the treatment such as pH, temperature, initial concentration of metal ions, dosage level of adsorbent etc. Adsorption process could also be controlled with types of mechanisms, like mass transfer, diffusion control, chemical reactions, and particle diffusion. Several adsorption models can be applied to evaluate the experimental data. In case of heavy metal adsorption by HAP or modified HAP, the pseudo-second-order model is best fitting at low concentrations while at higher concentration, the pseudo-first-order model is appropriately fitting the experimental data. The pseudo first-order equation is generally expressed as (Lagregren 1898).

$$\log(q_e - q_t) = \log(q_e) - \frac{k_1}{2.303}t$$

where q_e and q_t are the adsorption capacity at equilibrium and at time t, respectively (mg/g), k_1 is the rate constant of pseudo first-order adsorption (1/min). This is applicable in the initial stage of adsorption process.

The second-order Lagergren equation was given by Ho and McKay (1999).

$$\left(\frac{t}{q_t}\right) = \frac{l}{k_2 q_e^2} + \frac{l}{q_e}(t)$$

where k_2 is the rate constant of pseudo second-order adsorption (g/mg/min).

Adsorption isotherms help in assessing the optimal adsorption capacity and interactions of adsorbent. The most used adsorption isotherms are Langmuir and Freundlich for the sorbate-- adsorbent system. The Langmuir model is based on the assumption that binding sites have similar affinity for heavy metal ions, while in the Freundlich model it denotes heterogeneous systems in which adsorption sites have dissimilar affinities for sorbate species. Strontium-doped hydroxyapatite (Sr-HAP) adsorbent showed a maximum Cd(II) adsorption capacity (158.30 mg/g), followed by the Langmuir isothermal adsorption model and the pseudo-second-order kinetic model (Zhu et al., 2022). Kinetic properties of a sorbent depend on the occurrence of the active site and the ease of access of ions onto the active site without sterical hindrance. The mechanism of sorption will be subjected to the physical and chemical characteristics of the sorbent as well as on the mass transfer process.

6.5.2.2 REMOVAL OF DYES

Dyes are color materials which cannot be altered by different processes. There are natural dyes prepared from natural sources of flora and fauna while synthetic dyes are prepared from petroleum products. Health hazards associated with the dyes urge the necessity for a suitable dye removal process from the contaminated waters due to textile industry, the major causative. They are highly water soluble in nature and hence the process of removal of dye from water is complicated. HAP-based adsorbent can be used for removing anionic and cationic dyes (Pai et al., 2019). Mechanisms like coagulation-flocculation, biodegradation, electrochemical oxidation, Fenton oxidation, reverse osmosis, ozonation, electrocoagulation etc. are also involved in the dye removal process. Manatunga et al. (2018) reported removal of azo dye

AY220 using porous crystalline hydroxyapatite. Photocatalytic property of HAP is utilized in the removal of methyl mercaptan, dimethyl sulfide, and calgite. Removal of textile dyes by precipitation reaction at neutral pH by HAP was reported by Lemlikch et al. (2012). Removal of Congo red (CR) dye by adsorption using HAP nanoparticles as a fast and low-cost method is reported (Guan et al., 2018). Iron-doped HAP, cobalt-doped HAP, and bare HAP were compared for their efficacy in removing Congo red dye, highest removal efficiency was shown by iron-doped HAP at 95.6%. Polyalcohol-coated HAP showed effective adsorption of Congo red, methyl blue, and methyl orange (Guan et al., 2018). HAP, NaAlg-HAP, and MgHAP were evaluated for the removal of Acid Yellow 220 and in which NaAlg-HAP showed highest adsorption at 212.8 mg/g followed by HAP at 169.5 mg/g. HAP rapidly adsorbed methylene blue dye through electrostatic interaction and hydrogen bonding (Wei et al., 2015). Chitosan-doped HAP showed good adsorption capacity for Congo red removal at 769 mg/g. Similarly, chitosan-coated magnetic HAP was effective against reactive blue dye. The maximum for 50% weight of chitosan greater adsorption potential of nanocomposite forms of HAP with increased surface area is also documented (Pai et al., 2019). The adsorption of organic dye depends largely on the charge distribution of HAP molecule, with higher adsorption at pH values above and below isoelectric pH of HAP, that is between 6 and 7.2.

6.5.2.3 REMOVAL OF RADIONUCLIDES

Along with natural radionuclides, the occurrence of radioactive pollutants through anthropogenic sources poses ionizing radiation exposure which is a serious public health issue as it can damage, mutate, or even kill cells (Tran et al., 2021). Hence, it is critical to select an appropriate and effective method to remove radionuclides from water. The present methods involved are precipitation, reverse osmosis, ion exchange etc. which are expensive and involve maintenance. HAP can be used as an environment-friendly sorbent for radionuclides of uranium, neptunium, plutonium, americium, and strontium due to its poor water solubility, high chemical and thermal stability, high specific surface area, and good buffering properties (Shi et al., 2020; Galamboš et al., 2012). Cationic radionuclides replace Ca^{2+} in HAP. Hydroxyl groups of HAP primarily aid in the elimination of uranium by chelation (Guo et al., 2020). Industrial wastewater released from uranium mining and nuclear electrical power generation contain uranium ions which can lead to loss of environmental resources. Phosphate-derived materials effectively remove uranium from water by adsorption, which supports through surface complexation and mineralization of phosphate with uranium. HAP is efficient for use in permeable reactive barriers for uninterruptedly removing uranium (Chattanathan et al., 2013). Al-doped hydroxyapatite nanoparticles (Al-nHAP) expressed high adsorption against strontium (Ramakrishnan et al., 2016). The 90% of Sr^{2+} was adsorbed by the HAP column technique in the presence of competing cations and the adsorption increased with increase of pH (Nishiyama et al., 2016). Sorption of strontium and cesium ions from water using HAP cobalt(II) ferrocyanide showed enhanced removal by ion exchange mechanism (Metwally et al., 2017). Porous HAP beads synthesized by mixing HAP and polyvinyl butyral binder through sintering process is an effective adsorbent for removing radionuclides from aquatic environments (Tran 2021). Carbon nanotubes coated with strontium hydroxyapatite were successfully utilized in the removal of Co(II) and Eu(III) ions from aqueous solutions (Sheha et al., 2019).

6.5.2.4 REMOVAL OF FLUORIDE IONS

Elevated fluoride level in water due to natural and anthropogenic sources can pose many public health issues. World Health Organization specify a limit of 1.5 mg/L of fluoride in drinking water and the fluoride concentration varies in different geographical regions (WHO, 2006). HAP is illustrated as a suitable sorbent for the removal of fluoride from aqueous solutions with higher efficiency (Mourabet et al., 2015). The mechanisms involved in fluoride removal are ion-exchange and electrostatic interactions. He et al. (2016) reported the formation of stable fluorapatite $[Ca_5(PO4)_3F]$ and thermodynamically more stable mixed fluorinated HAP $[Ca_5(PO4)_3(OH, F)]$ by ion exchange. Positive ions of HAP bind to fluoride ions by adsorption through Vander Waals forces with enhanced interaction at lower pH.

At lower fluoride concentrations, the ion exchange reaction takes place by the following equations:

$$Ca_{10}(PO_4)_6(OH)_2(s) + F- \_aq\_ \rightarrow Ca_{10}(PO_4)_6(F,OH) \_aq\_ + OH- \_aq\_$$

$$Ca_{10}(PO_4)_6(F,OH)(s) + F- \_aq\_ \rightarrow Ca_{10}(PO_4)F2(s) + OH- \_aq\_$$

At higher concentrations, the surface phase of HAP is dissolved to form calcium fluoride as shown below:

$$Ca10(PO4\ )6(OH)2(s) + 20F- \_aq\_ \rightarrow 10CaF2(s)$$
$$+ 6PO3- 4 \_aq\_ + 2OH- \_aq\_$$

Important parameters influencing the fluoride removing mechanism of the HAP are contact time, dosage, particle size, solution pH, temperature, the presence of other anions and initial fluoride concentration (Mourabet et al., 2015). Nijhawan et al. (2017) prepared porous HAP ceramic beads of 4.0–5.0 mm using HAP and starch by the sintering method to adsorb the fluoride in column application as use of powdered HAP is impossible in continuous-flow column due to significant head loss. HAP containing multi-walled carbon nanotubes is an effective sorbent for fluoride ions. HAP modified activated alumina possesses a maximum adsorption capacity of 14.4 mg F/g which is five times higher than the activated alumina alone (Tomar et al., 2014).

6.5.2.5 REMOVAL OF EMERGING CONTAMINANTS

Highly stable compounds like cosmetics, herbicides, pesticides, plasticisers, hormones, and antibiotics do not easily degrade from the environment. Adsorption is an effective method for removing these pollutants.

6.5.2.5.1 Removal of Antimicrobials and Drugs

Antimicrobial resistance (AMR) has been emerged as a global health and development threat leading to the evolution of drug-resistant pathogens. The injudicious usage and disposal of antibiotics and other antimicrobials into natural waters are the major drivers in the development of AMR in human being. Ciprofloxacin, a broad-spectrum antibiotic of fluoroquinolone family, is an emerging pollutant regularly detected in natural waters. Adsorption on a porous matrix is an ideal mechanism for removing most of the antibiotics from water. Hydroxyapatite possess favorable physicochemical characteristics such as high chemical and thermal stability, low solubility in water, porous matrix structure, abundant reactive group, nontoxicity, green material, high regeneration capacity etc. Laabd et al. (2021) reported the effective removal of Ciprofloxacin from aqueous solutions using a composite of Hydroxyapatite and Montmorillonite synthesized by the coprecipitation method. Similarly, Oxytetracycline (OTC) is recorded as an emerging pollutant due to its serious impact on human health and the ecosystem. OTC is a broad-spectrum antibiotic, and hence is prescribed for a number of infections, therefore present in alarmingly high quantity in effluents from hospitals and animal farmlands. A hydroxyapatite composite with aluminosilicates prepared by the chemical precipitation method was reported as an efficient adsorbent material for the removal of OTC from wastewater (Martínez-Olivas et al., 2020). Similarly, the widely used fluoroquinolone antibiotics, norfloxacin (NOR) and ciprofloxacin (CIP), could be effectively adsorbed onto nano-hydroxyapatite (n-HAP) with a dose-dependent increase in adsorption efficiency. The adsorption was maximum at pH 6 (Chen et al., 2015). Ofloxacin and triclosan, which represent another frequent pollutant in waste water, were effectively adsorbed by HAP at an adsorption of triclosan (133.3 mg/g) and ofloxacin (29.15 mg/g) (Pai et al., 2020). The wood biochar doped with nano HAP was used to adsorb copper and tylosin/sulfamethoxazole through the formation of hydrogen bonding, π–π interactions, and complex formation. HAP-doped titania can degrade the pharmaceutical compounds like diclofenac and fluoxetine at a degradation capacity of more than 90% by incomplete mineralization (Marquez Brazon et al., 2016).

A nanocomposite adsorbent based on organophilic bentonite (OB), nano-hydroxyapatite (HA), and an ultra-high molecular weight polyethylene (UHMWPE) was fabricated with high adsorption efficiency toward tetracycline (Fagundes et al., 2019). The nanocomposite exhibited a maximum adsorption capacity of 8.72 mg/g for tetracycline at pH 7 within a contact time of 24 h. The major force of interaction reported was non-electrostatic interactions. Crosslinked amino hydroxyapatite/chitosan hybrids were fabricated by Pereira et al. (2020) for the elimination of diclofenac from aqueous solution. The composite exhibited an adsorption capacity of 125 mg/g for diclofenac drug. The drug was adsorbed onto the surface of hybrid primarily through electrostatic interaction. Chitosan /Al2O3-Hydroxyapatite nanocomposite beads were able to adsorb estradiol with maximum adsorption capacity exhibited by nano-composite beads found to be 39.78 mg/g (Li et al., 2020). Chemisorption was the major mechanism responsible behind the adsorption process.

6.5.2.5.2 Removal of Pesticides

The presence of pesticide pollutants in ground and surface water has become a major concern in agriculture intensified areas as it poses serious threat to human and animal health. Organophosphorus pesticides (OPPs) are the most commonly applied pesticide nowadays. Potassium substituted HAP and its magnetic nanocomposite KHA/Fe3O4 was reported as an effective adsorbent for the removal of organophosphorus pesticides (OPPs) from water samples (Chahkandi et al., 2019).

Various technologies have been performed for the removal of OPP pesticides from aqueous solutions such as filtration, coagulation, flocculation, sedimentation, adsorption, ion exchange, and distillation, of which advanced oxidation processes (AOPs) have been found to be most effective for the treatment of pesticide contaminated water. Out of various AOPs, photocatalysis has gained significant interest. Hydroxyapatite (HAP) possesses significant photocatalytic activity; it is a preferred catalytic agent on its own as in combination with other photocatalytic agents. Hydroxyapatite is preferred over other photocatalysts as it is having dual action as photocatalyst and as an efficient adsorbent. Chloropyrifos [O,O-diethyl-O-(3,5,6-trichloro-2-pyridinyl) phosphorothionate] (CP) is a crystalline OPP insecticide, which in human animal system hinders acetylcholinesterase enzyme activity and leads to improper functioning of the nervous system. Nanohydroxyapatite loaded with carboxyl functionalized graphene oxide/zinc oxide nanorod (nHAP@CFGO/ZnR) was demonstrated to have high photoactivity and photostability to degrade chlorpyrifos in aqueous solutions under visible light. The maximum adsorption occurs at pH 3.5. Similarly, HAP extracted from fish scale was found to have adsorption capacity toward bisphenol A (BPA), with a maximum adsorption capacity of 74.128 mg/g. About 84% of BPA could be eliminated at a dosage level of 1.375 g/L. Similarly, magnetic HAP is another attractive option as an adsorbent, because of the ease of recovery and reuse. Paraquat dichloride (PQ) is a broad-spectrum herbicide, frequently detected in natural waters. A nanocomposite of hydroxyapatite and magnetite was proposed as an efficient sorbent for PQ with 90% adsorption capacity at a dosage level of 416.67 mg/g within 130 min of contact time (Kotnala et al., 2022).

6.5.2.5.3 Removal of Nitrobenzene

Nitrobenzene, a carcinogenic pollutant, is widely used in the production of dyes, explosives, and pesticides, and is presently listed as a priority pollutant by the U.S. Environmental Protection Agency. Currently, activated carbon, silica zeolites, organo-clays, activated sludge etc. are used for the remediation of nitrobenzene in water, but are limited by several factors such as high cost, low adsorption capacity, poor mechanical strength, or narrow available pH range. Nanocrystalline hydroxyapatite HAP has been demonstrated as a rapid adsorbent for nitrobenzene from industrial effluents, which reached equilibrium within 1 min (Wei et al., 2010). Intensification of adsorbent concentration can increase effective surface area and sites of adsorption for the removal of pollutants. The removal efficiency of nitrobenzene increased from 66.2 to 70.9% with the increase in nano-crystalline HAP dosage from 2 to 5 g/L, and a further increase in dosage to 10 g/L enhanced the removal efficiency to 72.6% (Wei et al., 2010). Later on, hydroxyapatite–gelatin (HAP–GEL) nanocomposite was developed for nitrobenzene removal from aqueous solution with a maximum adsorption capacity of 42.373 mg/g (Wei et al., 2014). Higher adsorption capacity was observed at lower pH and temperature values.

6.5.2.5.4 Removal of Phenol

Phenol is increasingly used for a number of industries and is present in aquatic environment beyond permissible level. Phenol is conventionally removed from water using various adsorbents and the most widely employed adsorbent is activated carbon. However, activated carbon is relatively expensive and its regeneration is rather difficult. Nanosized HAP was found to be effective in removing phenol from waste water which reached equilibrium within 2 h of exposure. The absorption efficiency was maximum at acidic and alkaline pH ranges, the maximal adsorption capacity was observed at pH 2 (Lin et al., 2009). The mechanism of adsorption is both chemical and physical, and is primarily based on the hydrogen bonding between the hydrogen atom of phenol group and apatite metal ion or oxygen atom from the P–O group of hydroxyapatite molecule. The adsorption properties of HAP can be improved by surface modification. The surface modification of hydroxyapatite using alumina improves the phenol adsorption capacity of hydroxyapatite (Bouiahya et al., 2019). Narwade et al. (2017) reported a high phenol adsorption capacity for a composite film based on cellulose nanofibril and hydroxyapatite with a maximum adsorption capacity of 64 mg/g at pH 2. Phenol was adsorbed onto the surface of the composite film primarily by an intraparticle diffusion mechanism. The primary characteristics of HAP for using as an adsorbent are by virtue of its unique functionality, porosity, and the lattice structure. Owing to its high adsorption capacity toward phenol and related compounds, HAP can be very well explored for textile effluent remediation.

6.5.2.5.5 Removal of Nitrates and Nitrites

As agricultural and industrial activities are intensifying day by day, aquatic ecosystems are polluted with nitrite and nitrate compounds above safe limit.

As these nitrogen compounds are easily soluble in water, the removal of nitrogenous compounds is a difficult task. Conventionally, ion exchange, reverse osmosis, electrodialysis, catalytic hydrogenation, biological denitrification, adsorption etc. have been employed to eliminate nitrogenous products from drinking water. Among the various technologies, adsorption is found to be most cost-effective for the removal of nitrites and nitrates. Hydroxyapatite was demonstrated to be a suitable adsorbent material for the removal of nitrate as high as 96% adsorption efficiency under neutral conditions, at a dosage level of 0.3 g/100 mL 100 mg/L nitrate solution. The equilibrium was achieved rapidly within the first 40 min of treatment (Islam et al., 2010). HAP can be easily molded into compact, porous, and stable granules or beads with greater surface area. A novel type of magnetic nanosorbent composite of hydroxyapatite-coated Fe_2O_3 nanoparticles was experimented for the removal of nitrite and nitrate ions. Under the optimum conditions, the magnetic hydroxyapatite nanoparticles exhibited an adsorption efficiency of about 93–101% toward NO_3 − and NO_2 − ions (100 mg/L) (Ghasemi and Sillanpää, 2015).

6.5.3 ADVANTAGES OF NANO HYDROXYAPATITE (NANO HAP) IN WATER TREATMENT

The application of nano hydroxyapatite in wastewater treatment has been well documented in the literature. Cationic doping as well as organic molecule grafting of nano HAP encourage the immobilization of ions and enhance adsorption capacity. Nano HAP and its composites show better adsorption capacities on Cd^{2+}, Pb^{2+}, and Zn^{2+} compared to other materials (Balasooriya, et al., 2022) and heavy metal affinity of nano HAP is in the order of Pb^{2+}>Cd^{2+}> Ni^{2+} which is based on electro negativity of the ions and cation/anion state. The adsorption mechanism involved is mainly dissolution precipitation in which solubility is more in nano HAP, which depend upon the temperature and pH of the solution. Also, ionic exchange reaction and surface complexation with calcium, phosphate, and hydroxyl groups are involved. Among these mechanisms, dissolution–precipitation is the major mechanism in heavy metal removal by nano HAP. It dissolves in water and forms $H_2PO_4^-$ groups which precipitate heavy metal ions (Guo et al., 2017). The kinetic model that fits better in heavy metal removal is Lagergren pseudo-second, as per the literature.

Even though nano HAP is reported as an eco-friendly adsorbent of heavy metal ions, the agglomeration and precipitation nature of nano HAP reduce metal removal efficiency. Dispersion fixation of nano HAP in a polymer (with alginate fixation) was suggested by Guo et al. (2017) as a solution to this limitation. Composites of HAP with gelatin, chitosan, and chitin can impose a synergetic effect on adsorption capacity of composites, where nano HAP with dispersion in above bio polymers perform effectively in adsorption. Preparation of hybrid composite by polymer embedding is also reported as an alternative (Gopalakannan et al. 2018). Cu^{2+} reduction using chitin/chitosan nano HAP composite was reported by Gandhi et al. (2011) with high metal ion adsorption potential. Higher adsorption capacity was observed in chitosan nano HAP composite compared to chitin in combination and alone (Kousalya et al., 2010). Grafting of 1, 3-benzenedisulfonate on an HAP was found effective in the removal of Pb^{2+}, Zn^{2+}, and Cd^{2+} (Oulguidoum et al., 2021).

6.5.3.1 FACTORS INFLUENCING REMOVAL EFFICIENCY OF NANO HAP

The factors influencing adsorption efficiency of nano HAP are primarily, pH, dosage level, initial concentration, contact time, temperature, speed of agitation, and presence of other ions. At lower pH, competition of H^+ ions to metal ions leads to lower adsorption capacity, while at high pH, decrease of H^+ ions causes deprotonation of adsorbent resulting in the enhancement of metal ion adsorption (Balasooriya, et al., 2022). pH is considered crucial for phenol adsorption as well, as the removal of phenol by HAP is primarily through hydrogen bonding. A similar effect was observed in the case of nitrobenzene as well, whereas a better removal capacity for reactive red 198 dye was observed at acidic pH 6 (Vasugi et al., 2017). The adsorbent dosage determines the holding capacity of the sorbent for initial solute concentration which influences the increase of sorption sites and its availability. In case of low metal ions concentration, it could not reach the over-saturation point due to the excessive availability of adsorbents. Hence an appropriate dosage should be selected. The increase of adsorption capacity of nano HAP with the increase of metal concentration can be due to the increase in mass transfer driving force. The adsorption of heavy metals using HAP nanorods decreased progressively with the advent of time and reached equilibrium in 30 min (Mortada et al., 2015). An increasing trend of sorption with increase of temperature of water is also reported (Balasooriya, et al., 2022). The agitation speed of treatment can support the proper distribution of adsorbent in the medium. In multiple metal systems, selectivity of adsorbent is determined by the competitive adsorption of multiple metals on nano HAP. Chen et al. reported the sorption of nano HAP in adsorbing Cd, Pb, and Cu, and observed a higher preference for Pb followed by Cu and Cd. The ratios of competitive metal ions displacement are moderately consistent over varied concentrations in multiple metal systems.

6.5.4 RECENT TRENDS IN THE APPLICATION OF HAP IN WATER TREATMENT

Application of HAP nanorods and nanocomposites showed efficient lead removal (Mohammad et al., 2017). Sricharoen et al. (2020) reported the application of Nile tilapia fish scale derived HAP along with ultrasonic action for effective removal of Hg^{2+}. Fish bone derived HAP was observed as a remediating agent for removing U(VI) radionuclides from aqueous solutions in a laboratory batch mode (Lammers et al., 2017; Han et al., 2018). HAP disc is fabricated as a filtration device for waste water (Hariyanto et al., 2016). Microporous hydroxyapatite ceramic beads for fluoride removal from drinking water showed high efficiency in removal (Nijhawan et al., 2017). 3D printed water purification filters based on polylactic acid (PLA) functionalized with HAP from fish scale for heavy metal removal were reported by Fijoł et al. (2021). Ultralong hydroxyapatite nanowires and cellulose fibers have been developed for high-performance water purification (Zhang et al., 2019). Magnetic hydroxyapatite-immobilized oxidized multi-walled carbon nanotubes are reported as an excellent adsorbent for Pb(II) and methylene blue (Wang et al., 2017). As the separation of nano-adsorbent for harmful hazards after environmental release is a potential issue and can aggravate the condition to worsen, magnetic recycling can prevent the entry of adsorbed hazards into the environment.

6.6 CONCLUSION

The expense of water purification can be minimized by using cost-effective and renewable adsorbents which demand less energy. The current chapter is a systematic presentation of hydroxyapatite, outlining the sources, synthesis protocols, and the application of HAP for water purification. Emphasis has been given to biological apatite (bioapatite) from aquatic resources. HAPs can be extracted from renewable resources such as fish-processing discards in a cost-effective manner and can be easily regenerated and reused. The various mechanisms of pollutant removal by HAP such as heavy metals, synthetic dyes, and other emerging pollutants are detailed citing documented literature.

Compared to microcrystalline HAP, nano HAp and its composites are having high adsorption efficiency especially toward heavy metals. Among the various parameters, pH is a crucial factor that determines the adsorption efficiency of HAP as the major mechanism controlling adsorption of HAP is electrostatic interaction. Magnetic nano HAP can overcome the major disadvantages of native HAP, as magnetic HAP can be easily recovered and regenerated from the solution using external magnetic field. It can be concluded that hydroxyapatite is a candidate adsorbent material for the large-scale treatment of waste water for the removal of heavy metal ions, radionuclides, organic pollutants, fluoride ions, and several other industrial contaminants in wastewater. Research efforts need to be intensified in developing HAP-based treatment devices specific to the individual applications. HAP extracted from renewable sources using green synthesis protocols are highly recommended for the future.

KEYWORDS

- **hydroxyapatite**
- **aquatic source**
- **water treatment**
- **adsorption**

REFERENCES

1. Alif, M. F.; Aprillia, W.; Arief, S. A Hydrothermal Synthesis of Natural Hydroxyapatite Obtained from *Corbicula moltkiana* Freshwater Clams Shell Biowaste. *Mater. Lett.* **2018,** *230*, 40–43.
2. Anandan, A.; Janakiram, T. Adsorption of Cr (VI) from Aqueous Solutions by Low Cost Activated Carbon. *J. Chem. Pharm. Res.* **2012,** *4* (6), 2900–2905.
3. Anirudhan, T. S.; Shainy, F.; Sekhar, V. C.; Athira, V. S. Highly Efficient Photocatalytic Degradation of Chlorpyrifos in Aqueous Solutions by Nano Hydroxyapatite Modified CFGO/ZnO Nanorod Composite. *J. Photochem. Photobiol. A* **2021,** *418*, 113333.
4. Ashraf, S.; Ahmed, M. K.; Ibrahium, H. A.; Awwad, N. S.; Abdel-Fattah, E.; Ghoniem, M. G. Nanofibers of Polycaprolactone Containing Hydroxyapatite Doped with Aluminum/Vanadate Ions for Wound Healing Applications. *New J. Chem.* **2021,** *45* (48), 22610–22620.
5. Athinarayanan, J.; Periasamy, V. S. and Alshatwi, A. A. Simultaneous Fabrication of Carbon Nanodots and Hydroxyapatite Nanoparticles from Fish Scale for Biomedical Applications. *Mater. Sci. Eng. C* **2020,** *117*, 111313.

6. Aziz, K.; Mamouni, R.; Azrrar, A.; Kjidaa, B.; Saffaj, N.; Aziz, F. Enhanced Biosorption of Bisphenol A from Wastewater Using Hydroxyapatite Elaborated from Fish Scales and Camel Bone Meal: A RSM@ BBD Optimization Approach. *Ceram. Int.* **2022,** *48* (11), 15811–15823.
7. Balasooriya, I. L.; Chen, J.; Korale Gedara, S. M.; Han, Y.; Wickramaratne, M. N. Applications of Nano Hydroxyapatite as Adsorbents: A Review. *Nanomaterials* **2022,** *12* (14), 2324.
8. Bano, N.; Jikan, S. S.; Basri, H.; Bakar, S. A. A.; Nuhu, A. H. Natural Hydroxyapatite Extracted from Bovine Bone. *J. Sci. Technol.* **2017,** *9* (2).
9. Binsi, P. K.; Ashraf, P. M.; Thomas, N. G.; Zynudheen, A. A. Thermal Transition Analysis of Fish Scale Originated Hydroxyapatite: AFM and Electrochemical Impedance Spectroscopy as Complimentary Techniques. *J. Thermal Anal. Calorimetry* **2022,** *147* (6), 4027–4045.
10. Bouiahya, K.; Es-saidi, I.; El Bekkali, C.; Laghzizil, A.; Robert, D.; Nunzi, J. M.; Saoiabi, A. Synthesis and Properties of Alumina-Hydroxyapatite Composites from Natural Phosphate for Phenol Removal from Water. *Colloid Interface Sci. Commun.* **2019,** *31*, 100188.
11. Boutinguiza, M.; Pou, J.; Comesaña, R.; Lusquiños, F.; De Carlos, A.; León, B. Biological Hydroxyapatite Obtained from Fish Bones. *Mater. Sci. Eng. C* **2012,** *32* (3), 478–486.
12. Brazdis, R. I.; Fierascu, I.; Avramescu, S. M.; Fierascu, R. C. Recent Progress in the Application of Hydroxyapatite for the Adsorption of Heavy Metals from Water Matrices. *Materials* **2021,** *14* (22), 6898.
13. Bunsiri, R.; Thamaphat, K.; Limsuwan, P. Synthesis and Characterization of Pure Natural Hydroxyapatite from Fish Bones Bio-Waste. In *Advanced Materials Research*, Vol. 506; Trans Tech Publications Ltd, 2012; pp. 206–209.
14. Campisi, S.; Castellano, C.; Gervasini, A. Tailoring the Structural and Morphological Properties of Hydroxyapatite Materials to Enhance the Capture Efficiency Towards Copper (II) and Lead (II) Ions. *New J. Chem.* **2018,** *42* (6), 4520–4530.
15. Chahkandi, M.; Amiri, A.; Arami, S. R. S. Extraction and Preconcentration of Organophosphorus Pesticides from Water Samples and Fruit Juices Utilizing Hydroxyapatite/Fe3O4 Nanocomposite. *Microchem. J.* **2019,** *144*, 261–269.
16. Chattanathan, S. A.; Clement, T. P.; Kanel, S. R.; Barnett, M. O.; Chatakondi, N. Remediation of Uranium-Contaminated Groundwater by Sorption Onto Hydroxyapatite Derived from Catfish Bones. *Water Air Soil Pollut.* **2013,** *224* (2), 1–9.
17. Chaudon, L.; Cecille, L.; Klein, M.; Kowa, S.; Mehling, O.; Thiels, G. Evaluation of Alternative Waste Management Schemes for LWR Hulls and Caps (No. CEA-CONF—10454). CEA Etablissement de la Vallee du Rhone, 1990.
18. Chen, L.; Zhang, K. S.; He, J. Y.; Xu, W. H.; Huang, X. J.; Liu, J. H. Enhanced Fluoride Removal from Water by Sulfate-Doped Hydroxyapatite Hierarchical Hollow Microspheres. *Chem. Eng. J.* **2016,** *285*, 616–624.
19. Chen, Y.; Lan, T.; Duan, L.; Wang, F.; Zhao, B.; Zhang, S.; Wei, W. Adsorptive Removal and Adsorption Kinetics of Fluoroquinolone by Nano-Hydroxyapatite. *PlOS One* **2015,** *10* (12), e0145025.
20. Choi, J. M.; Kong, Y. M.; Kim, S.; Kim, H. E.; Hwang, C. S.; Lee, I. S. Formation and Characterization of Hydroxyapatite Coating Layer on Ti-Based Metal Implant by Electron-Beam Deposition. *J. Mater. Res.* **1999,** *14* (7), pp.2980–2985.
21. Ciobanu, G.; Barna, S.; Harja, M. Kinetic and Equilibrium Studies on Adsorption of Reactive Blue 19 Dye from Aqueous Solutions by Nanohydroxyapatite Adsorbent. *Arch. Environ. Prot.* **2016,** *42*, 3–11.

22. Corami, A.; Mignardi, S.; Ferrini, V. Copper and Zinc Decontamination from Single-and Binary-Metal Solutions Using Hydroxyapatite. *J. Hazard. Mater.* **2007,** *146* (1–2), 164–170.
23. Da Cruz, J. A.; Weinand, W. R.; Neto, A. M.; Palácios, R. S.; Sales, A. J. M.; Prezas, P. R.; Costa, M. M.; Graça, M. P. F. Low-Cost Hydroxyapatite Powders from Tilapia Fish. *JOM* **2020,** *72* (4), 1435–1442.
24. Dabiri, S. M. H.; Rezaie, A. A.; Moghimi, M.; Rezaie, H. Extraction of Hydroxyapatite from Fish Bones and Its Application in Nickel Adsorption. *BioNanoSci.* **2018,** *8* (3), 823–834.
25. Das, M. P.; Vijaylakshmi, J. V.; Sugunaw, J. V.; Renuka, M.; Prasad, K. Efficient Dye Decolorization of an Azo Dye on Fish Scale Hydroxyapatite. *Res. J. Pharm. Technol.* **2019,** *12* (6), 2917–2921.
26. Deng, F.; Luo, X. B.; Ding, L.; Luo, S. L. Application of Nanomaterials and Nanotechnology in the Reutilization of Metal Ion from Wastewater. In *Nanomaterials for the Removal of Pollutants and Resource Reutilization*; Elsevier, 2019; pp 149–178.
27. Eswaran, M.; Swamiappan, S.; Chokkiah, B.; Dhanusuraman, R.; Bharathkumar, S.; Ponnusamy, V. K. A Green and Economical Approach to Derive Nanostructured Hydroxyapatite from Garra Mullya Fish Scale Waste for Biocompatible Energy Storage Applications. *Mater. Lett.* **2021,** *302*, 130341.
28. Fagundes, A. P.; da Silva, A. F. V.; Macuvele, D. L. P.; Padoin, N.; Soares, C.; Riella, H. G. Ultra-High-Molecular-Weight Polyethylene-Based Nanocomposite for Removal of Tetracycline in Aqueous Systems. *J. Environ. Chem. Eng.* **2020,** *8* (3), 103630.
29. Fagundes, A. P.; Macuvele, D. L. P.; Padoin, N.; Soares, C.; Riella, H. G. A Novel Ultrahigh-Molecular-Weight Polyethylene-Based Nanocomposite for Contaminants Adsorption in Aqueous Systems. *Mater. Lett.* **2019,** *240*, 197–200.
30. Faksawat, K.; Sujinnapram, S.; Limsuwan, P.; Hoonnivathana, E.; Naemchanthara, K. Preparation and Characteristic of Hydroxyapatite Synthesized from Cuttlefish Bone by Precipitation Method. In *Advanced Materials Research*, Vol. **1125**; Trans Tech Publications Ltd, 2015; pp 421–425.
31. Ferri, M.; Campisi, S.; Scavini, M.; Evangelisti, C.; Carniti, P.; Gervasini, A. In-Depth Study of the Mechanism of Heavy Metal Trapping on the Surface of Hydroxyapatite. *Appl. Surf. Sci.* **2019,** *475*, 397–409.
32. Fijoł, N.; Abdelhamid, H. N.; Pillai, B.; Hall, S. A.; Thomas, N.; Mathew, A. P. 3D-Printed Monolithic Biofilters Based on a Polylactic Acid (PLA)–Hydroxyapatite (HAp) Composite for Heavy Metal Removal from an Aqueous Medium. *RSC Adv.* **2021,** *11* (51), 32408–32418.
33. Foroutan, R.; Peighambardoust, S. J.; Hosseini, S. S.; Akbari, A.; Ramavandi, B. Hydroxyapatite Biomaterial Production from Chicken (Femur and Beak) and Fishbone Waste Through a Chemical Less Method for Cd2+ Removal from Shipbuilding Wastewater. *J. Hazard. Mater.* **2021,** *413*, 125428.
34. Freundlich, H. Über die Adsorption in Lösungen. Z. Phys. Chem.; 57U, 385–470.
35. Galamboš, M.; Suchánek, P.; Rosskopfová, O. Sorption of Anthropogenic Radionuclides on Natural and Synthetic Inorganic Sorbents. *J. Radioanal. Nucl. Chem.* **2012,** *293* (2), 613–633.
36. Gandhi, M. R.; Kousalya, G. N.; Meenakshi, S. Removal of Copper (II) Using Chitin/Chitosan Nano-Hydroxyapatite Composite. *Int. J. Biol. Macromol.* **2011,** *48* (1), 119–124.
37. Gao, S.; Cui, J.; Wei, Z. Study on the Fluoride Adsorption of Various Apatite Materials in Aqueous Solution. *J. Fluorine Chem.* **2009,** *130* (11), 1035–1041.

38. George, S.; Mehta, D.; Saharan, V. K. Application of Hydroxyapatite and Its Modified Forms as Adsorbents for Water Defluoridation: An Insight Into Process Synthesis. *Rev. Chem. Eng.* **2020,** *36* (3), 369–400.
39. Ghasemi, E.; Sillanpää, M. Magnetic Hydroxyapatite Nanoparticles: An Efficient Adsorbent for the Separation and Removal of Nitrate and Nitrite Ions from Environmental Samples. *J. Sep. Sci.* **2015,** *38* (1), 164–169.
40. Gopalakannan, V.; Periyasamy, S.; Viswanathan, N. Fabrication of Magnetic Particles Reinforced Nano-Hydroxyapatite/Gelatin Composite for Selective Cr (VI) Removal from Water. *Environ. Sci.: Water Res. Technol.* **2018,** *4* (6), 783–794.
41. Guan, Y.; Cao, W.; Guan, H.; Lei, X.; Wang, X.; Tu, Y. et al. A Novel Polyalcohol-Coated Hydroxyapatite for the Fast Adsorption of Organic Dyes. *Colloids Surf. A* **2018,** *548*, 85–91.
42. Guan, Y.; Cao, W.; Wang, X.; Marchetti, A.; Tu, Y. Hydroxyapatite Nano-Rods for the Fast Removal of Congo Red Dye from Aqueous Solution. *Mater. Res. Express* **2018,** *5* (6), 065053.
43. Guo, J.; Han, Y.; Mao, Y.; Wickramaratne, M. N. Influence of Alginate Fixation on the Adsorption Capacity of Hydroxyapatite Nanocrystals to Cu2+ Ions. *Colloids Surf. A* **2017,** *529*, 801–807.
44. Guo, Y.; Gong, Z.; Li, C.; Gao, B.; Li, P.; Wang, X. et al. Efficient Removal of Uranium (VI) by 3D Hierarchical Mg/Fe-LDH Supported Nanoscale Hydroxyapatite: A Synthetic Experimental and Mechanism Studies. *Chem. Eng. J.* **2020,** *392*, 123682.
45. Gupta, V. K.; Nayak, A.; Bhushan, B.; Agarwal, S. A Critical Analysis on the Efficiency of Activated Carbons from Low-Cost Precursors for Heavy Metals Remediation. *Crit. Rev. Environ. Sci. Technol.* **2015,** *45* (6), 613–668.
46. Hamzah, S.; Yatim, N. I.; Alias, M.; Ali, A.; Rasit, N.; Abuhabib, A. Extraction of Hydroxyapatite from Fish Scales and Its Integration with Rice Husk for Ammonia Removal in Aquaculture Wastewater. *Indonesian J. Chem.* **2019,** *19* (4), 1019–1030.
47. Han, M.; Kong, L.; Hu, X.; Chen, D.; Xiong, X.; Zhang, H. et al. Phase Migration and Transformation of Uranium in Mineralized Immobilization by Wasted Bio-Hydroxyapatite. *J. Clean. Product.* **2018,** *197*, 886–894.
48. Handley-Sidhu, S.; Renshaw, J. C.; Yong, P.; Kerley, R.; Macaskie, L. E. Nano-Crystalline Hydroxyapatite Bio-Mineral for the Treatment of Strontium from Aqueous Solutions. *Biotechnol. Lett.* **2011,** 33(1), 79–87.
49. Hariani, P. L.; Muryati, M.; Said, M.; Salni, S. Synthesis of Nano-Hydroxyapatite from Snakehead (*Channa striata*) Fish Bone and Its Antibacterial Properties. In *Key Engineering Materials*, Vol. 840; Trans Tech Publications Ltd, 2020; pp 293–299.
50. Hariyanto, S. D.; Sriani, T.; Prihandana, G. S. Fabrication of Hydroxyapatite Disc as Filtration Membrane in Waste Water Treatment System. In *Applied Mechanics and Materials*, Vol. 842; Trans Tech Publications Ltd., 2016; pp 461–466.
51. Harja, M.; Ciobanu, G. Studies on Adsorption of Oxytetracycline from Aqueous Solutions onto Hydroxyapatite. *Sci. Total Environ.* **2018,** *628*, 36–43.
52. Hassan, A. F.; Hrdina, R. Chitosan/Nanohydroxyapatite Composite Based Scallop Shells as an Efficient Adsorbent for Mercuric Ions: Static and Dynamic Adsorption Studies. *Int. J. Biol. Macromol.* **2018,** *109*, 507–516.
53. Hassan, M. N.; Mahmoud, M. M.; Abd El-Fattah, A. and Kandil, S. Microwave-Assisted Preparation of Nano-Hydroxyapatite for Bone Substitutes. *Ceram. Int.* **2016,** *42* (3), 3725–3744.

54. He, J.; Zhang, K.; Wu, S.; Cai, X.; Chen, K.; Li, Y. et al. Performance of Novel Hydroxyapatite Nanowires in Treatment of Fluoride Contaminated Water. *J. Hazard. Mater.* **2016,** *303*, 119–130.
55. Herawati, R. A.; Faisal, M. Utilization of Hydroxyapatite from Tuna Fish Bone Waste as Adsorbent for Cadmium from Aqueous Solutions. In *IOP Conference Series: Materials Science and Engineering*, Vol. 845, No. 1; IOP Publishing, 2020; p 012025.
56. Hernández-Cocoletzi, H.; Salinas, R. A.; Águila-Almanza, E.; Rubio-Rosas, E.; Chai, W. S.; Chew, K. W.; Mariscal-Hernández, C.; Show, P. L. Natural Hydroxyapatite from Fishbone Waste for the Rapid Adsorption of Heavy Metals of Aqueous Effluent. *Environ. Technol. Innov.* **2020,** *20*, 101109.
57. Ho, Y. S.; McKay, G. Pseudo-Second-Order Model for Sorption Processes. *Process Biochem.* **1999,** *34*, 451–465.
58. Hou, H.; Zhou, R.; Wu, P.; Wu, L. Removal of Congo Red Dye from Aqueous Solution with Hydroxyapatite/Chitosan Composite. *Chem. Eng. J.* **2012,** *211*, 336–342.
59. Huang, Y. C.; Chu, H. W. Using Hydroxyapatite from Fish Scales to Prepare Chitosan/Gelatin/Hydroxyapatite Membrane: Exploring Potential for Bone Tissue Engineering. *J. Marine Sci. Technol.* **2013,** *21* (6), 13.
60. Ibrahim, M.; Labaki, M.; Giraudon, J. M.; Lamonier, J. F. Hydroxyapatite, a Multifunctional Material for Air, Water and Soil Pollution Control: A Review. *J. Hazard. Mater.* **2020,** *383*, 121139.
61. Ibrahim, R. B. Contaminated Land Management and Control Guideline s No. 1: Malaysian Recommended Site Screening Levels for Contaminate d Land. Kuala Lumpur, Malaysia, 2009.
62. Islam, M.; Mishra, P. C.; Patel, R. Physicochemical Characterization of Hydroxyapatite and Its Application Towards Removal of Nitrate from Water. *J. Environ. Manag.* **2010,** *91* (9), 1883–1891.
63. Jayaweera, H. D. A.; Siriwardane, I.; de Silva, K. M.; de Silva, R. M. Synthesis of Multifunctional Activated Carbon Nanocomposite Comprising Biocompatible Flake Nano Hydroxyapatite and Natural Turmeric Extract for the Removal of Bacteria and Lead Ions from Aqueous Solution. *Chem. Central J.* **2018,** *12* (1), 1–14.
64. Karacan, I.; Cox, N.; Dowd, A.; Vago, R.; Milthorpe, B.; Cazalbou, S.; Ben-Nissan, B. The Synthesis of Hydroxyapatite from Artificially Grown Red Sea Hydrozoan Coral for Antimicrobacterial Drug Delivery System Applications. *J. Aust. Ceram. Soc.* **2021,** *57* (2), 399–407.
65. Kazeminezhad, I.; Ahmadizadeh, S.; Babaie, A. A. Application of Magnetic Hydroxyapatite Nanoparticles for Removal of Cd2+ from Aqueous Solutions, 2014.
66. Khanday, W. A.; Hameed, B. H. Zeolite-Hydroxyapatite-Activated Oil Palm Ash Composite for Antibiotic Tetracycline Adsorption. *Fuel* **2018,** *215*, 499–505.
67. Kim, J.; Sambudi, N. S.; Cho, K. Removal of Sr2+ Using High-Surface-Area Hydroxyapatite Synthesized by Non-Additive In Situ Precipitation. *J. Environ. Manag.* **2019,** *231*, 788–794.
68. Kotnala, S.; Bhushan, B.; Nayak, A. Fabrication of a Magnetite Hydroxyapatite Nanocomposite for the Removal of Paraquat Dichloride: Adsorption Studies. *Mater. Today: Proc.* **2022.**
69. Kousalya, G. N.; Gandhi, M. R.; Sundaram, C. S.; Meenakshi, S. Synthesis of Nano-Hydroxyapatite Chitin/Chitosan Hybrid Biocomposites for the Removal of Fe (III). *Carbohydr. Polym.* **2010,** *82* (3), 594–599.

70. Krestou, A.; Xenidis, A.; Panias, D. Mechanism of Aqueous Uranium (VI) Uptake by Hydroxyapatite. *Miner. Eng.* **2004,** *17* (3), 373–381.
71. Krishnamurithy, G.; Murali, M. R.; Hamdi, M.; Abbas, A. A.; Raghavendran, H. B. and Kamarul, T. Characterization of Bovine-Derived Porous Hydroxyapatite Scaffold and Its Potential to Support Osteogenic Differentiation of Human Bone Marrow Derived Mesenchymal Stem Cells. *Ceram. Int.* **2014,** *40* (1), 771–777.
72. Laabd, M.; Brahmi, Y.; El Ibrahimi, B.; Hsini, A.; Toufik, E.; Abdellaoui, Y.; Abou Oualid, H.; El Ouardi, M.; Albourine, A. A Novel Mesoporous Hydroxyapatite@ Montmorillonite Hybrid Composite for High-Performance Removal of Emerging Ciprofloxacin Antibiotic from Water: Integrated Experimental and Monte Carlo Computational Assessment. *J. Mol. Liquids* **2021,** *338*, 116705.
73. Lagregren, S. About the Theory of So-Called Adsorption of Soluble Substances. *Kungliga Svenska Vetenskapsakademiens Handlingar* **1989,** *24*, 1–39.
74. Lammers, L. N.; Rasmussen, H.; Adilman, D.; deLemos, J. L.; Zeeb, P.; Larson, D. G.; Quicksall, A. N. Groundwater Uranium Stabilization by a Metastable Hydroxyapatite. *Appl. Geochem.* **2017,** *84*, 105–113.
75. Langmuir, I. The Adsorption of Gases on Plane Surfaces of Glass, Mica and Platinum. *J. Am. Chem. Soc.* **1918,** *40*, 1361–1403.
76. Le Ho, K. H.; Dao, V. H.; Pham, X. K.; Nguyen, P. A.; Phan, B. V.; Doan, T. T.; Lam, T. H. Physicochemical Properties, Acute and Subchronic Toxicity of Nano-Hydroxyapatite Obtained from Lates Calcarifer Fish Bone. *Region. Stud. Marine Sci.* **2022,** *55*, 102560.
77. Lemlikchi, W.; Sharrock, P.; Mecherri, M. O.; Fiallo, M.; Nzihou, A. Treatment of Textile Waste Waters by Hydroxyapatite Co-precipitation with Adsorbent Regeneration and Reuse. *Waste Biomass Valorization* **2012,** *3* (1), 75–79.
78. Li, L.; Iqbal, J.; Zhu, Y.; Wang, F.; Zhang, F.; Chen, W. et al. Chitosan/Al2O3-HA Nanocomposite Beads for Efficient Removal of Estradiol and Chrysoidin from Aqueous Solution. *Int. J. Biol. Macromol.* **2020,** *145*, 686–693.
79. Li, L.; Iqbal, J.; Zhu, Y.; Zhang, P.; Chen, W.; Bhatnagar, A.; Du, Y. Chitosan/Ag-Hydroxyapatite Nanocomposite Beads as a Potential Adsorbent for the Efficient Removal of Toxic Aquatic Pollutants. *Int. J. Biol. Macromol.* **2018,** *120*, 1752–1759.
80. Li, Y.; Wang, S.; Zhang, Y.; Han, R.; Wei, W. Enhanced Tetracycline Adsorption onto Hydroxyapatite by Fe (III) Incorporation. *J. Mol. Liquids* **2017,** *247*, 171–181.
81. Liaw, B. S.; Chang, T. T.; Chang, H. K.; Liu, W. K.; Chen, P. Y. Fish Scale-Extracted Hydroxyapatite/Chitosan Composite Scaffolds Fabricated by Freeze Casting—An Innovative Strategy for Water Treatment. *J. Hazard. Mater.* **2020,** *382*, 121082.
82. Lin, K.; Pan, J.; Chen, Y.; Cheng, R.; Xu, X. Study the Adsorption of Phenol from Aqueous Solution on Hydroxyapatite Nanopowders. *J. Hazard. Mater.* **2009,** *161* (1), 231–240.
83. Liu, W. K.; Liaw, B. S.; Chang, H. K.; Wang, Y. F.; Chen, P. Y. From Waste to Health: Synthesis of Hydroxyapatite Scaffolds from Fish Scales for Lead Ion Removal. *JOM* **2017,** *69* (4), 713–718.
84. Liu, Y.; Liu, M.; Ji, S.; Zhang, L.; Cao, W.; Wang, H.; Wang, S. Preparation and Application of Hydroxyapatite Extracted from Fish Scale Waste Using Deep Eutectic Solvents. *Ceram. Int.* **2021,** *47* (7), 9366–9372.
85. Maghsoodi, M. R.; Ghodszad, L.; Lajayer, B. A. Dilemma of Hydroxyapatite Nanoparticles as Phosphorus Fertilizer: Potentials, Challenges and Effects on Plants. *Environ. Technol. Innov.* **2020,** *19*, 100869.

86. Manatunga, D. C.; de Silva, R. M.; Nalin de Silva, K. M.; de Silva, N.; Premalal, E. V. A. Metal and Polymer-Mediated Synthesis of Porous Crystalline Hydroxyapatite Nanocomposites for Environmental Remediation. *R. Soc. Open Sci.* **2018,** *5* (1), 171557.
87. Marquez Brazon, E.; Piccirillo, C.; Moreira, I. S.; Castro, P. M. L. Photodegradation of Pharmaceutical Persistent Pollutants Using Hydroxyapatite-Based Materials. *J. Environ. Manage.* **2016,** *182*, 486–495.
88. Martínez-Olivas, A.; Torres-Pérez, J.; Balderas-Hernández, P.; Reyes-López, S. Y. Oxytetracycline Sorption onto Synthetized Materials from Hydroxyapatite and Aluminosilicates. *Water Air Soil Pollut.* **2020,** *231* (6), 1–21.
89. Mehta, D.; Mondal, P.; Saharan, V. K.; George, S. In Vitro Synthesis of Marble Apatite as a Novel Adsorbent for Removal of Fluoride Ions from Ground Water: An Ultrasonic Approach. *Ultrason. Sonochem.* **2018,** *40*, 664–674.
90. Metwally, S. S.; Ahmed, I. M.; Rizk, H. E. Modification of Hydroxyapatite for Removal of Cesium and Strontium Ions from Aqueous Solution. *J. Alloys Compd* **2017,** *709*, 438–444.
91. Mirhosseini, M.; Biazar, E.; Saeb, K. Removal of Arsenic from Drinking Water by Hydroxyapatite Nano Particles. *Curr. World Environ.* **2014,** *9* (2), 331.
92. Mobasherpour, I.; Salahi, E.; Pazouki, M. Comparative of the Removal of Pb2+, Cd2+ and Ni2+ by Nano Crystallite Hydroxyapatite from Aqueous Solutions: Adsorption Isotherm Study. *Arab. J. Chem.* **2012,** *5* (4), 439–446.
93. Mohammad, A. M.; Eldin, T. A. S.; Hassan, M. A.; El-Anadouli, B. E. Efficient Treatment of Lead-Containing Wastewater by Hydroxyapatite/Chitosan Nanostructures. *Arab. J. Chem.* **2017,** *10* (5), 683–690.
94. Moore, R. C.; Hasan, A. A. M.; Holt, K. C.; Hasan, M. A. Overview on Backfill Materials and Permeable Reactive Barriers for Nuclear Waste Disposal Facilities (No. SAND2003–2101). Sandia National Laboratories (SNL), Albuquerque, NM, and Livermore, CA (United States), 2003.
95. Mortada, W. I.; Kenawy, I. M.; Abdelghany, A. M.; Ismail, A. M.; Donia, A. F.; Nabieh, K. A. Determination of Cu^{2+}, Zn^{2+} and Pb^{2+} in Biological and Food Samples by FAAS After Preconcentration with Hydroxyapatite Nanorods Originated from Eggshell. *Mater. Sci. Eng. C* **2015,** *52*, 288–296.
96. Mourabet, M.; El Rhilassi, A.; El Boujaady, H.; Bennani-Ziatni, M.; El Hamri, R.; Taitai, A. Removal of Fluoride from Aqueous Solution by Adsorption on Hydroxyapatite (HAp) Using Response Surface Methodology. *J. Saudi Chem. Soc.* **2015,** *19* (6), 603–615.
97. Muhammad, N.; Gao, Y.; Iqbal, F.; Ahmad, P.; Ge, R.; Nishan, U.; Rahim, A.; Gonfa, G.; Ullah, Z. Extraction of Biocompatible Hydroxyapatite from Fish Scales Using Novel Approach of Ionic Liquid Pretreatment. *Sep. Purif. Technol.* **2016,** *161*, 129–135.
98. Mustafa, N.; Ibrahim, M. H. I.; Asmawi, R.; Amin, A. M. Hydroxyapatite Extracted from Waste Fish Bones and Scales via Calcination Method. In *Appl. Mech.* Mater., Vol. 773; Trans Tech Publications Ltd, 2015; pp 287–290.
99. Nandi, S. K.; Kundu, B.; Mukherjee, J.; Mahato, A.; Datta, S.; Balla, V. K. Converted Marine Coral Hydroxyapatite Implants with Growth Factors: In Vivo Bone Regeneration. *Mater. Sci. Eng. C* **2015,** *49*, 816–823.
100. Narwade, V. N.; Anjum, S. R.; Kokol, V.; Khairnar, R. S. Ammonia-Sensing Ability of Differently Structured Hydroxyapatite Blended Cellulose Nanofibril Composite Films. *Cellulose* **2019,** *26* (5), 3325–3337.
101. Narwade, V. N.; Khairnar, R. S.; Kokol, V. In Situ Synthesised Hydroxyapatite-Loaded Films Based on Cellulose Nanofibrils for Phenol Removal from Wastewater. *Cellulose* **2017,** *24* (11), 4911–4925.

102. Nguyen, V. C.; Pho, Q. H. Preparation of Chitosan Coated Magnetic Hydroxyapatite Nanoparticles and Application for Adsorption of Reactive Blue 19 and Ni2+ Ions. *Sci. World J.* **2014,** *2014*.
103. Nijhawan, A.; Butler, E. C.; Sabatini, D. A. Macroporous Hydroxyapatite Ceramic Beads for Fluoride Removal from Drinking Water. *J. Chem. Technol. Biotechnol.* **2017,** *92* (8), 1868–1875.
104. Nishiyama, Y.; Hanafusa, T.; Yamashita, J.; Yamamoto, Y.; Ono, T. Adsorption and Removal of Strontium in Aqueous Solution by Synthetic Hydroxyapatite. *J. Radioanal. Nucl. Chem.* **2016,** *307* (2), 1279–1285.
105. Opre, Z.; Grunwaldt, J. D.; Maciejewski, M.; Ferri, D.; Mallat, T.; Baiker, A. Promoted Ru–Hydroxyapatite: Designed Structure for the Fast and Highly Selective Oxidation of Alcohols with Oxygen. *J. Catal.* **2005,** *230* (2), 406–419.
106. Oulguidoum, A.; Bouiahya, K.; Bouyarmane, H.; Talbaoui, A.; Nunzi, J. M.; Laghzizil, A. Mesoporous Nanocrystalline Sulfonated Hydroxyapatites Enhance Heavy Metal Removal and Antimicrobial Activity. *Sep. Purif. Technol.* **2021,** *255*, 117777.
107. Ozawa, M.; Suzuki, S. Microstructural Development of Natural Hydroxyapatite Originated from Fish-Bone Waste Through Heat Treatment. *J. Am. Ceram. Soc.* **2002,** *85* (5), 1315–1317.
108. Pai, S.; Kini, M. S.; Selvaraj, R. A Review on Adsorptive Removal of Dyes from Wastewater by Hydroxyapatite Nanocomposites. *Environ. Sci. Pollut. Res.* **2021,** *28* (10), 11835–11849.
109. Pai, S.; Kini, S. M.; Selvaraj, R.; Pugazhendhi, A. A Review on the Synthesis of Hydroxyapatite, Its Composites and Adsorptive Removal of Pollutants from Wastewater. *J. Water Process Eng.* **2020,** *38*, 101574.
110. Pal, A.; Paul, S.; Choudhury, A. R.; Balla, V. K.; Das, M.; Sinha, A. Synthesis of Hydroxyapatite from Lates Calcarifer Fish Bone for Biomedical Applications. *Mater. Lett.* **2017,** *203*, 89–92.
111. Pandi, K.; Viswanathan, N. Synthesis of Alginate Bioencapsulated Nano-Hydroxyapatite Composite for Selective Fluoride Sorption. *Carbohydr. Polym.* **2014,** *112*, 662–667.
112. Pandi, K.; Viswanathan, N. Enhanced Defluoridation and Facile Separation of Magnetic Nano-Hydroxyapatite/Alginate Composite. *Int. J. Biol. Macromol.* **2015,** *80*, 341–349.
113. Periyasamy, S.; Gopalakannan, V.; Viswanathan, N. Hydrothermal Assisted Magnetic Nano-Hydroxyapatite Encapsulated Alginate Beads for Efficient Cr (VI) Uptake from Water. *J. Environ. Chem. Eng.* **2018,** *6* (1), 1443–1454.
114. Piccirillo, C.; Pereira, S. I. A.; Marques, A. P. G. C.; Pullar, R. C.; Tobaldi, D. M.; Pintado, M. E.; Castro, P. M. L. Bacteria Immobilisation on Hydroxyapatite Surface for Heavy Metals Removal. *J. Environ. Manage.* **2013,** *121*, 87–95.
115. Prabakaran, K.; Rajeswari, S. Development of Hydroxyapatite from Natural Fish Bone Through Heat Treatment. *Trends Biomater. Artif. Organs* **2006,** *20* (1), 20–23.
116. Rajec, P.; Mátel, L.; Orechovská, J.; Šúcha, J.; Novák, I. Sorption of Radionuclides on Inorganic Sorbents. *J. Radioanal. Nucl. Chem.* **1996,** *208* (2), 477–486.
117. Ramakrishnan, P.; Nagarajan, S.; Thiruvenkatam, V.; Palanisami, T.; Naidu, R.; Mallavarapu, M.; Rajendran, S. Cation Doped Hydroxyapatite Nanoparticles Enhance Strontium Adsorption from Aqueous System: A Comparative Study with and Without Calcination. *Appl. Clay Sci.* **2016,** *134*, 136–144.
118. Robinson, J. S.; Baumann, K.; Hu, Y.; Hagemann, P.; Kebelmann, L.; Leinweber, P. Phosphorus Transformations in Plant-Based and Bio-Waste Materials Induced by Pyrolysis. *Ambio* **2018,** *47* (1), 73–82.

119. Ruan, Z.; Tian, Y.; Ruan, J.; Cui, G.; Iqbal, K.; Iqbal, A. et al. Synthesis of Hydroxyapatite/Multi-Walled Carbon Nanotubes for the Removal of Fluoride Ions from Solution. *Appl. Surf. Sci.* **2017,** *412*, 578–590.
120. Rubi, R. V.; Roque, E.; Rosa, F. D.; Estoque, R. M.; Olvido, G.; Perey, P. J. et al. Photocatalytic Degradation of Atrazine Herbicide Using Nano-Hydroxyapatite from Cow Bone synthesized via Simulated Body Fluid. In *IOP Conference Series: Materials Science and Engineering*, Vol. 778, No. 1; IOP Publishing, 2020; p. 012013.
121. Samant, A.; Nayak, B.; Misra, P. K. Kinetics and Mechanistic Interpretation of Fluoride Removal by Nanocrystalline Hydroxyapatite Derived from *Limacine artica* Shells. *J. Environ. Chem. Eng.* **2017,** *5* (6), 5429–5438.
122. Sánchez-Bayo, F. Sources and Toxicity of Pollutants. In: *Ecological Impacts of Toxic Chemicals Chapter: 1*; Sánchez-Bayo, F.; van den Brink, P. J.; Mann, R. M., Eds.; Bentham Science Publishers, 2011; pp 3–12.
123. Sasaki, K.; Goto, T. Immobilization of Sr2+ on Naturally Derived Hydroxyapatite by Calcination of Different Species of Fish Bones and Influence of Calcination on Ion-Exchange Efficiency. *Ceram. Int.* **2014,** *40* (8), 11649–11656.
124. Selimin, M. A.; Latif, A. F. A.; Lee, C. W.; Muhamad, M. S.; Basri, H.; Lee, T. C. Adsorption Efficiency of Hydroxyapatite Synthesised from Black Tilapia Fish Scales for Chromium (VI) Removal. *Mater. Today: Proc.* **2022,** *57*, 1142–1146.
125. Sheha, R. R.; Moussa, S. I.; Attia, M. A.; Sadeek, S. A.; Someda, H. H. Development and Application of Carbon Nanotubes Reinforced Hydroxyapatite Composite in Separation of Co (II) and Eu (III) Ions from Aqueous Solutions. *Radiochim. Acta* **2019,** *107* (1), 67–82.
126. Shi, P.; Liu, M.; Fan, F.; Yu, C.; Lu, W.; Du, M. Characterization of Natural Hydroxyapatite Originated from Fish Bone and Its Biocompatibility with Osteoblasts. *Mater. Sci. Eng. C* **2018,** *90*, 706–712.
127. Shi, S.; Li, B.; Qian, Y.; Mei, P.; Wang, N. A Simple and Universal Strategy to Construct Robust and Anti-Biofouling Amidoxime Aerogels for Enhanced Uranium Extraction from Seawater. *Chem. Eng. J.* **2020,** *397*, 125337.
128. Silvester, L.; Lamonier, J. F.; Vannier, R. N.; Lamonier, C.; Capron, M.; Mamede, A. S. et al. Structural, Textural and Acid–Base Properties of Carbonate-Containing Hydroxyapatites. *J. Mater. Chem. A* **2014,** *2* (29), 11073–11090.
129. Singh, S.; Khare, A.; Chaudhari, S. Enhanced Fluoride Removal from Drinking Water Using Non-Calcined Synthetic Hydroxyapatite. *J. Environ. Chem. Eng.* **2020,** *8* (2), 103704.
130. Sricharoen, P.; Kongsri, S.; Kukusamude, C.; Areerob, Y.; Nuengmatcha, P.; Chanthai, S.; Limchoowong, N. Ultrasound-Irradiated Synthesis of 3-Mercaptopropyl Trimethoxysilane-Modified Hydroxyapatite Derived from Fish-Scale Residues Followed by Ultrasound-Assisted Organic Dyes Removal. *Sci. Rep.* **2021,** *11* (1), 1–11.
131. Sricharoen, P.; Limchoowong, N.; Nuengmatcha, P.; and Chanthai, S. Ultrasonic-Assisted Recycling of Nile Tilapia Fish Scale Biowaste into Low-Cost Nano-Hydroxyapatite: Ultrasonic-Assisted Adsorption for Hg2+ Removal from Aqueous Solution Followed by "Turn-Off" Fluorescent Sensor Based on Hg2+-Graphene Quantum Dots. *Ultrason. Sonochem.* **2020,** *63*, 104966.
132. Stötzel, C.; Müller, F. A.; Reinert, F.; Niederdraenk, F.; Barralet, J. E.; Gbureck, U. Ion Adsorption Behaviour of Hydroxyapatite with Different Crystallinities. *Colloids Surf. B* **2009,** *74* (1), 91–95.

133. Strunz, H.; Nickel, E. H. *Strunz Mineralogical Tables: Chemical–Structural Mineral Classification System*, 9th ed.; Schweizerbart: Stuttgart, 2001.
134. Sunil, B. R.; Jagannatham, M. Producing Hydroxyapatite from Fish Bones by Heat Treatment. *Mater. Let.* **2016,** *185*, 411–414.
135. Surya, P.; Nithin, A.; Sundaramanickam, A.; Sathish, M. Synthesis and Characterization of Nano-Hydroxyapatite from *Sardinella longiceps* Fish Bone and Its Effects on Human Osteoblast Bone Cells. J. Mech. Behav. Biomed. Mater. **2021,** *119*, 104501.
136. Tomar, G.; Thareja, A.; Sarkar, S. Fluoride Removal by a Hybrid Fluoride-Selective Adsorbent. *Water Sci. Technol. Water Supply* **2014,** *14* (6), 1133–1141.
137. Tomar, G.; Thareja, A.; Sarkar, S. Enhanced Fluoride Removal by Hydroxyapatite-Modified Activated Alumina. *Int. J. Environ. Sci. Technol.* **2015,** *12* (9), 2809–2818.
138. Tran, T. N.; Kim, J.; Park, J. S.; Chung, Y.; Han, J.; Oh, S.; Kang, S. Novel Hydroxyapatite Beads for the Adsorption of Radionuclides from Decommissioned Nuclear Power Plant Sites. *Appl. Sci.* **2021,** *11* (4), 1746.
139. Vasugi, G.; Thamizhavel, A.; Girija, E. K. Surface Modification of Hydroxyapatite by Yttrium Ion Substitution for Adsorption of Reactive Red Dye. *J. Mater. Environ. Sci.* **2017,** *8*, 714–723.
140. Venkatesan, J.; Kim, S. K. Effect of Temperature on Isolation and Characterization of Hydroxyapatite from Tuna (*Thunnus obesus*) Bone. *Materials* **2010,** *3* (10), 4761–4772
141. Venkatesan, J.; Lowe, B.; Manivasagan, P.; Kang, K.-H.; Chalisserry, E. P.; Anil, S.; Kim, D. G.; Kim, S.-K. Isolation and Characterization of Nano-Hydroxyapatite from Salmon Fish Bone. *Materials* **2015,** *8* (8), 5426–5439.
142. Venkatesan, J.; Rekha, P. D.; Anil, S.; Bhatnagar, I.; Sudha, P. N.; Dechsakulwatana, C.; Kim, S.-K.; Shim, M. S. Hydroxyapatite from Cuttlefish Bone: Isolation, Characterizations, and Applications. *Biotechnol. Bioprocess Eng.* **2018,** *23* (4) 383–393.
143. Vila, M.; Sánchez-Salcedo, S.; Cicuéndez, M.; Izquierdo-Barba, I.; Vallet-Regí, M. Novel Biopolymer-Coated Hydroxyapatite Foams for Removing Heavy-Metals from Polluted Water. *J. Hazard. Mater.* **2011,** *192* (1), 71–77.
144. Wakamura, M.; Kandori, K.; Ishikawa, T. Surface structure and composition of calcium hydroxyapatites substituted with Al (III), La (III) and Fe (III) ions. *Colloids and Surfaces A: Physicochemical and Engineering Aspects*, *164*(2–3), 297–305.
145. Wang, Y.; Hu, L.; Zhang, G.; Yan, T.; Yan, L.; Wei, Q.; Du, B. Removal of Pb (II) and Methylene Blue from Aqueous Solution by Magnetic Hydroxyapatite-Immobilized Oxidized Multi-Walled Carbon Nanotubes. *J. Colloid Interface Sci.* **2017,** *494*, 380–388.
146. Wardani, S. C.; Sujuti, H.; Mustamsir, E.; Hapsari, D. N. Synthesis and Potential of Skipjack Tuna Bone Hydroxyapatite as Bone Tissue Engineering Biomaterial. *J. Phys.* **2020,** *1665* (1), 012032.
147. Wei, W.; Sun, R.; Cui, J.; Wei, Z. Removal of Nitrobenzene from Aqueous Solution by Adsorption on Nanocrystalline Hydroxyapatite. *Desalination* **2010,** *263* (1–3), 89–96.
148. Wei, W.; Sun, R.; Jin, Z.; Cui, J.; Wei, Z. Hydroxyapatite–Gelatin Nanocomposite as a Novel Adsorbent for Nitrobenzene Removal from Aqueous Solution. *Appl. Surf. Sci.* **2014,** *292*, 1020–1029.
149. Wei, W.; Yang, L.; Zhong, W. H.; Li, S. Y.; Cui, J.; Wei, Z. G. Fast Removal of Methylene Blue from Aqueous Solution by Adsorption Onto Poorly Crystalline Hydroxyapatite Nanoparticles. *Dig. J. Nanomater. Biostruct.* **2015,** *19*, 1343–1363.
150. World Health Organization. Guidelines for Drinking-Water Quality [Electronic Resource]: Incorporating First Addendum, Vol. 1, Recommendations, 2006.

151. Yin, G.; Liu, Z.; Zhan, J.; Ding, F.; Yuan, N. Impacts of the Surface Charge Property on Protein Adsorption on Hydroxyapatite. *Chem. Eng. J.* **2002,** *87* (2), 181–186.
152. Zhang, Q. Q.; Zhu, Y. J.; Wu, J.; Shao, Y. T.; Cai, A. Y.; Dong, L. Y. Ultralong Hydroxyapatite Nanowire-Based Filter Paper for High-Performance Water Purification. *ACS Appl. Mater. Interfaces* **2019,** *11* (4), 4288–4301.
153. Zhang, Z.; Wang, X.; Wang, H.; Zhao, J. Removal of Pb (II) from Aqueous Solution Using Hydroxyapatite/Calcium Silicate Hydrate (HAP/CSH) Composite Adsorbent Prepared by a Phosphate Recovery Process. *Chem. Eng. J.* **2018,** *344*, 53–61.
154. Zhou, Y.; Li, S.; Wang, D.; Han, X. Electrospinning Synthesis of Hydroxyapatite Nanofibers Assembled from Nanorods and Their Adsorption for Heavy Metal Ions. *Polish J. Environ. Stud.* **2019,** *28* (2).
155. Zhu, Z.; Yang, Y.; Fan, Y.; Zhang, L.; Tang, S.; Zhu, Y.; Zhou, X. Strontium-Doped Hydroxyapatite as an Efficient Adsorbent for Cd (II) Removal from Wastewater: Performance, Kinetics, and Mechanism. *Environ. Technol. Innov.* **2022,** *28*, 102575.

Index

X